Arduino III

Internet of Things

Synthesis Lectures on Digital Circuits and Systems

Editor
Mitchell A. Thornton, *Southern Methodist University*

The *Synthesis Lectures on Digital Circuits and Systems* series is comprised of 50- to 100-page books targeted for audience members with a wide-ranging background. The Lectures include topics that are of interest to students, professionals, and researchers in the area of design and analysis of digital circuits and systems. Each Lecture is self-contained and focuses on the background information required to understand the subject matter and practical case studies that illustrate applications. The format of a Lecture is structured such that each will be devoted to a specific topic in digital circuits and systems rather than a larger overview of several topics such as that found in a comprehensive handbook. The Lectures cover both well-established areas as well as newly developed or emerging material in digital circuits and systems design and analysis.

Arduino III: Internet of Things
Steven F. Barrett
2021

Arduino II: Systems
Steven F. Barrett
2020

Arduino I: Getting Started
Steven F. Barrett
2020

Index Generation Functions
Tsutomu Sasao
2019

Microchip AVR® Microcontroller Primer: Programming and Interfacing, Third Edition
Steven F. Barrett and Daniel J. Pack
2019

Microcontroller Programming and Interfacing with Texas Instruments MSP430FR2433 and MSP430FR5994 – Part II, Second Edition
Steven F. Barrett and Daniel J. Pack
2019

Arduino III: Internet of Things

Steven F. Barrett

ISBN: 978-3-031-79922-8   paperback
ISBN: 978-3-031-79923-5   ebook
ISBN: 978-3-031-79924-2   hardcover

DOI 10.1007/978-3-031-79923-5

A Publication in the Springer series
SYNTHESIS LECTURES ON DIGITAL CIRCUITS AND SYSTEMS

Lecture #60
Series Editor: Mitchell A. Thornton, *Southern Methodist University*
Series ISSN
Print 1932-3166   Electronic 1932-3174

Arduino III

Internet of Things

Steven F. Barrett
University of Wyoming, Laramie, WY

SYNTHESIS LECTURES ON DIGITAL CIRCUITS AND SYSTEMS #60

ABSTRACT

This book is about the Arduino microcontroller and the Arduino concept. The visionary Arduino team of Massimo Banzi, David Cuartielles, Tom Igoe, Gianluca Martino, and David Mellis launched a new innovation in microcontroller hardware in 2005, the concept of open-source hardware. Their approach was to openly share details of microcontroller-based hardware design platforms to stimulate the sharing of ideas and promote innovation. This concept has been popular in the software world for many years. In June 2019, Joel Claypool and I met to plan the fourth edition of *Arduino Microcontroller Processing for Everyone!* Our goal has been to provide an accessible book on the rapidly evolving world of Arduino for a wide variety of audiences including students of the fine arts, middle and senior high school students, engineering design students, and practicing scientists and engineers. To make the book even more accessible to better serve our readers, we decided to change our approach and provide a series of smaller volumes. Each volume is written to a specific audience. This book, *Arduino III: Internet of Things*, explores Arduino applications in the fascinating and rapidly evolving world of the Internet of Things. *Arduino I: Getting Started* provides an introduction to the Arduino concept. *Arduino II: Systems*, is a detailed treatment of the ATmega328 processor and an introduction to C programming and microcontroller-based systems design.

KEYWORDS

Arduino microcontroller, Arduino UNO R3, Internet of Things, IoT, MKR 1000, MKR1010, greenhouse, weather stat

Contents

Preface

This book is about the Arduino microcontroller and the Arduino concept. The visionary Arduino team of Massimo Banzi, David Cuartielles, Tom Igoe, Gianluca Martino, and David Mellis launched a new innovation in microcontroller hardware in 2005, the concept of open-source hardware. Their approach was to openly share details of microcontroller-based hardware design platforms to stimulate the sharing of ideas and promote innovation. This concept has been popular in the software world for many years. In June 2019, Joel Claypool and I met to plan the fourth edition of *Arduino Microcontroller Processing for Everyone!* Our goal has been to provide an accessible book on the rapidly evolving world of Arduino for a wide variety of audiences including students of the fine arts, middle and senior high school students, engineering design students, and practicing scientists and engineers. To make the book even more accessible to better serve our readers, we decided to change our approach and provide a series of smaller volumes. Each volume is written to a specific audience. This book, *Arduino III: Internet of Things*, explores Arduino applications in the fascinating and rapidly evolving world of the Internet of Things (IoT). *Arduino I: Getting Started* provides an introduction to the Arduino concept. *Arduino II: Systems* is a detailed treatment of the ATmega328 processor and an introduction to C programming and microcontroller-based systems design.

APPROACH OF THE BOOK

The goal of this book series is to provide a thorough introduction to the Arduino UNO R3 and interfacing to different peripherals, a detailed exploration of the ATmega328, and the embedded systems design process, and Arduino applications in the IoT using the Arduino MKR 1000 and MKR 1010, as shown in Figure 1. We try very hard to strike a good balance between theory and practical information. The theory is important to understand some of the underlying concepts. We also think it is important to provide multiple, practical examples they may be adapted for other projects.

This book, *Arduino III: the Internet of Things*, explores Arduino applications in the IoT. The phrase "Internet of Things" or "IoT" is attributed to Kevin Ashton in a 1999 presentation describing supply chain initiatives [Hanes].[1] Generally speaking, IoT is about connecting objects via the internet to accomplish specific tasks. A closely related concept is cyber-physical systems (CPS) or the integration of computer and physical processes. Although IoT and CPS originated in different communities, they share many of the same techniques and concepts. A National Institute of Standards and Technology (NIST) study acknowledges the "emerging

---

[1] Hanes D., G. Salgueiro, P. Grossetete, R. Barton, J. Henry (2017) *IoT Fundamentals–Networking Technologies, Protocols, and Use Cases for the Internet of Things*, Cisco Press.

Arduino I: Getting Started	Arduino II: Systems	Arduino III: Internet of Things
Getting started	Getting started	Getting started
Arduino platforms	Programming	Internet
Power and interfacing	Analog-to-digital conversion	Internet of Things (IoT)
System examples	Timing subsystem	Connectivity
	Serial communication subsystem	Application: Greenhouse
	Interrupt subsystem	
	Embedded systems design	

Figure 1: Arduino book series.

consensus around the equivalence of CPS and IoT concepts [Greer]."[2] We do the same in this book and primarily use the term "IoT" to refer to both.

The book builds upon the foundation of the first two volumes. The reader should have a solid grounding in the Arduino UNO R3, the Arduino Development Environment, and writing Arduino sketches. For completeness we have provided prerequisite information in Chapter 1. Chapter 1 provides a brief review of some of Arduino concepts and introduces the Microchip ATmega328. This is the processor hosted onboard the Arduino UNO R3.

Chapter 2 provides an introduction to the fascinating world of the internet. In planning this chapter, a concept map was constructed of the internet. The concept map allows one to diagram and provide organization and relationship between related concepts. It is a good tool to take a large concept, such as the internet, and break it into smaller, related topics. From the concept map, the chapter outline was developed. The concept map of the internet developed for Chapter 2 is provided in Figure 2. The concept map is not unique or complete. It may evolve over time to take into account emerging concepts.

Chapter 2 also provides an introduction to the IoT and CPS. It provides a brief introduction to the IoT and CPS concepts and then describes enabling technologies, a simplified IoT architecture, the related topics of Information Technology (IT), and Operational Technology (OT), and explores the Industrial Internet of Things (IIoT). IIoT is a rapidly emerging concept in many areas of agriculture, energy, food production, manufacturing, mining, transportation, and many other industries (Hanes [5]).

Chapter 3 explores the multiple methods a microcontroller may be connected to another device to share information and tasks. The chapter begins with an overview of techniques based on range and a tutorial on serial communication concepts. Specific connection techniques are then discussed including Near Field Communication (NFC); short-range, hardwire, serial communication systems (UART, SPI, I2C); and radio frequency (RF) connection techniques including BlueTooth, ZigBee, Ethernet, and connection via the cellular phone network.

---

[2]Greer, C., M. Burns, D. Wollman, E. Griffor (2019) *Cyber–Physical Systems and Internet of Things*, NIST Special Publications 1900–202, National Institute of Standards and Technology, U.S. Department of Commerce.

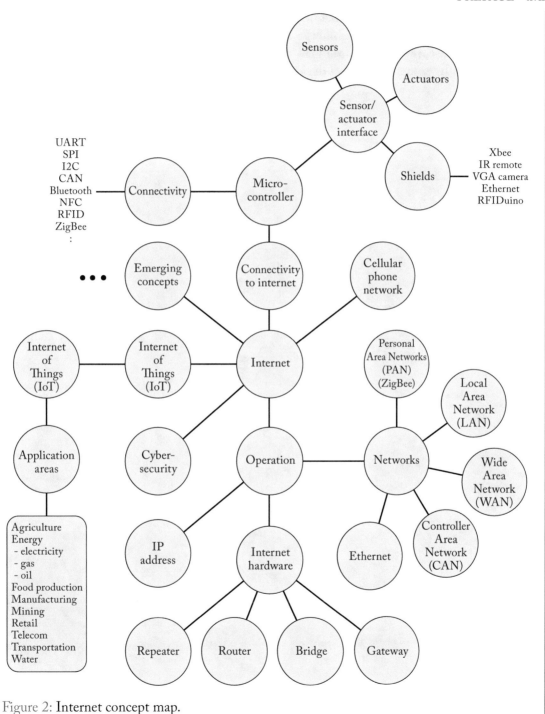

Figure 2: Internet concept map.

Chapter 4 applies the information from the previous chapters in the design and construction of an instrumented greenhouse. Although the application is specifically geared toward a greenhouse, the multiple examples provided of equipping an Arduino UNO R3 and a MKR 1000 with sensors and control devices may be applied to a wide variety of projects.

Steven F. Barrett
Laramie, Wyoming, March 2021

Acknowledgments

A number of people have made this book possible. I would like to thank Massimo Banzi of the Arduino design team for his support and encouragement in writing the first edition of this book: *Arduino Microcontroller: Processing for Everyone!* In 2005, Joel Claypool of Morgan & Claypool Publishers, invited Daniel Pack and I to write a book on microcontrollers for his new series titled "Synthesis Lectures on Digital Circuits and Systems." The result was the book *Microcontrollers Fundamentals for Engineers and Scientists.* Since then we have been regular contributors to the series. Our goal has been to provide the fundamental concepts common to all microcontrollers and then apply the concepts to the specific microcontroller under discussion. We believe that once you have mastered these fundamental concepts, they are easily transportable to different processors. As with many other projects, Joel has provided his publishing expertise to convert our final draft into a finished product. We thank him for his support on this project and many others. He has provided many novice writers the opportunity to become published authors. His vision and expertise in the publishing world made this book possible. We also thank Dr. C.L. Tondo of T&T TechWorks, Inc. and his staff for working their magic to convert our final draft into a beautiful book.

I would also like to thank Sparkfun, Adafruit, DFRobot, and Microchip for their permission to use images of their products and copyrighted material throughout the text series. Several Microchip acknowledgments are in order.

- This book contains copyrighted material of Microchip Technology Incorporated replicated with permission. All rights reserved. No further replications may be made without Microchip Technology Inc.'s prior written consent.

- *Arduino III: Internet of Things* is an independent publication and is not affiliated with, nor has it been authorized, sponsored, or otherwise approved by Microchip.

I would like to dedicate this book to my close friend Dr. Daniel Pack, Ph.D., P.E. In 2000, Daniel suggested that we might write a book together on microcontrollers. I had always wanted to write a book but I thought that's what other people did. With Daniel's encouragement we wrote that first book (and several more since then). Daniel is a good father, good son, good husband, brilliant engineer, great leader, has a work ethic that is second to none, and is a good friend. To you, good friend, I dedicate this book. I know that we will write many more together.

It is hard to believe we have been writing together for 20+ years. Finally, I would like to thank my wife and best friend of many years, Cindy.

Steven F. Barrett
Laramie, Wyoming, March 2021

CHAPTER 1

Getting Started

Objectives: After reading this chapter, the reader should be able to do the following.

- Successfully download and execute a simple program using the Arduino Development Environment (ADE).

- Describe the key features of the ADE.

- Name and describe the different features aboard the Arduino UNO R3 processor board.

- Discuss the features and functions of the Microchip ATmega328.

1.1   OVERVIEW

Welcome to the world of Arduino![1] The Arduino concept of open-source hardware was developed by the visionary Arduino team of Massimo Banzi, David Cuartilles, Tom Igoe, Gianluca Martino, and David Mellis in Ivrea, Italy. The team's goal was to develop a line of easy-to-use microcontroller hardware and software such that processing power would be readily available to everyone.

We assume you have a solid footing in the Arduino UNO R3, the ADE, interfacing techniques, and Arduino sketch writing. The chapter begins with a brief review of some of these concepts.

We use a top-down design approach. We begin with the "big picture" of the chapter. We then discuss the ADE and how it may be used to quickly develop a program (sketch) for the Arduino UNO R3. We then provide an overview of the hardware features of the Arduino UNO R3 evaluation board which hosts the Microchip ATmega328 processor.

1.2   THE BIG PICTURE

Most microcontrollers are programmed with some variant of the C programming language. The C programming language provides a nice balance between the programmer's control of the microcontroller hardware and time efficiency in program writing. As an alternative, the ADE provides a user-friendly interface to quickly develop a program, transform the program

---

[1]This chapter is included with permission from *Arduino I: Getting Started* for completeness and to allow each series volume to be independent.

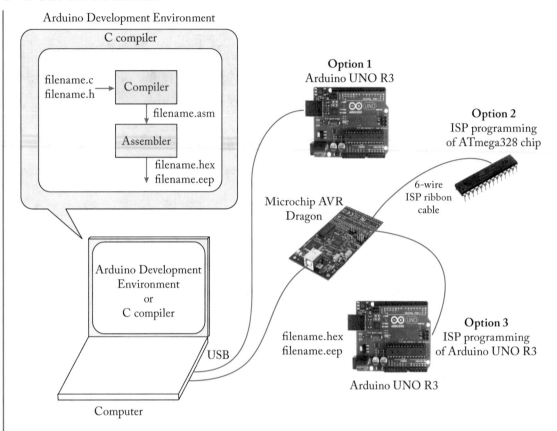

Figure 1.1: Programming the Arduino processor board. (Arduino illustrations used with permission of the Arduino Team (CC BY-NC-SA) [www.arduino.cc]. Microchip AVR Dragon illustration used with permission of Microchip, Incorporated [www.microchip.com].)

to machine code, and then load the machine code into the Arduino processor in several simple steps, as shown in Figure 1.1.

The first version of the ADE was released in August 2005. It was developed at the Interaction Design Institute in Ivrea, Italy to allow students the ability to quickly put processing power to use in a wide variety of projects. Since that time, updated versions incorporating new features have been released on a regular basis [www.arduino.cc].

At its most fundamental level, the ADE is a user-friendly interface to allow one to quickly write, load, and execute code on a microcontroller. A barebones program need only consist of a setup() and loop() function. The ADE adds the other required pieces such as header files and the main program construct. The ADE is written in Java and has its origins in the Processor programming language and the Wiring Project [www.arduino.cc].

The ADE is hosted on a laptop or personal computer (PC). Once the Arduino program, referred to as a sketch, is written; it is verified and uploaded to the Arduino UNO R3 evaluation board. Alternatively, a program may be written in C using a compiler. The compiled code can be uploaded to the Arduino UNO R3 using a programming pod such as the Microchip AVR Dragon.

1.3   ARDUINO QUICK START

To get started using an Arduino-based platform, you will need the following hardware and software:

- an Arduino-based hardware processing platform,

- the appropriate interface cable from the host PC or laptop to the Arduino platform,

- an Arduino compatible power supply, and

- the Arduino software.

Interface cable.   The Arduino UNO R3 connects to the host laptop or PC via a USB cable (type A male to type B female).

Power supply.   The Arduino processing boards may be powered from the USB port during project development. However, it is highly recommended that an external power supply be employed. This will allow developing projects beyond the limited electrical current capability of the USB port. For the UNO R3 platform, Arduino [www.arduino.cc] recommends a power supply from 7–12 VDC with a 2.1 mm center positive plug. A power supply of this type is readily available from a number of electronic parts supply companies. For example, the Jameco #133891 power supply is a 9 VDC model rated at 300 mA and equipped with a 2.1 mm center positive plug. It is available for under US$10. The UNO has an onboard voltage regulators that maintain the incoming power supply voltage to a stable 5 VDC.

1.3.1   QUICK START GUIDE

The ADE may be downloaded from the Arduino website's front page at www.arduino.cc. Versions are available for Windows, Mac OS X, and Linux. Provided below is a quick start step-by-step approach to blink an onboard LED.

- Download the ADE from www.arduino.cc.

- Connect the Arduino UNO R3 processing board to the host computer via a USB cable (A male to B male).

- Start the ADE.

- Under the Tools tab select the type of evaluation **Board** you are using and the **Port** that it is connected to.

- Type the following program.

```
//****************************************************************
#define LED_PIN 13

void setup()
{
pinMode(LED_PIN, OUTPUT);
}

void loop()
{
digitalWrite(LED_PIN, HIGH);
delay(500);                      //delay specified in ms
digitalWrite(LED_PIN, LOW);
delay(500);
}
//****************************************************************
```

- Upload and execute the program by asserting the "Upload" (right arrow) button.

- The onboard LED should blink at one second intervals.

With the ADE downloaded and exercised, let's take a closer look at its features.

1.3.2 ARDUINO DEVELOPMENT ENVIRONMENT OVERVIEW

The ADE is illustrated in Figure 1.2. The ADE contains a text editor, a message area for displaying status, a text console, a tool bar of common functions, and an extensive menuing system. The ADE also provides a user-friendly interface to the Arduino processor board which allows for a quick upload of code. This is possible because the Arduino processing boards are equipped with a bootloader program.

A close-up of the Arduino toolbar is provided in Figure 1.3. The toolbar provides single button access to the more commonly used menu features. Most of the features are self-explanatory. As described in the previous section, the "Upload" button compiles your code and uploads it to the Arduino processing board. The "Serial Monitor" button opens the serial monitor feature. The serial monitor feature allows text data to be sent to and received from the Arduino processing board.

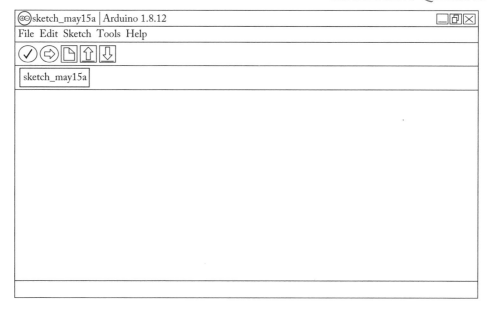

Figure 1.2: Arduino Development Environment [www.arduino.cc].

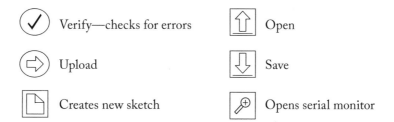

Figure 1.3: Arduino Development Environment buttons.

1.3.3   SKETCHBOOK CONCEPT

In keeping with a hardware and software platform for students of the arts, the Arduino environment employs the concept of a sketchbook. An artist maintains their works in progress in a sketchbook. Similarly, programs are maintained within a sketchbook in the Arduino environment. Furthermore, we refer to individual programs as sketches. An individual sketch within the sketchbook may be accessed via the Sketchbook entry under the File tab.

1.3.4   ARDUINO SOFTWARE, LIBRARIES, AND LANGUAGE REFERENCES

The ADE has a number of built-in features. Some of the features may be directly accessed via the ADE drop-down toolbar illustrated in Figure 1.2. Provided in Figure 1.4 is a handy reference to

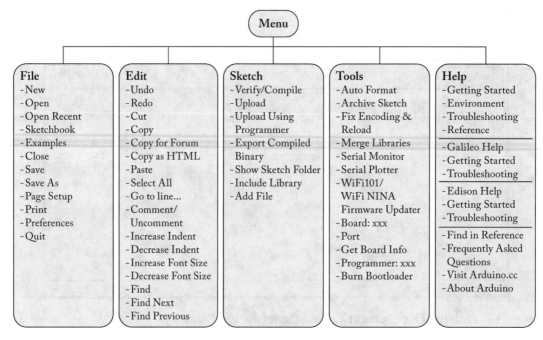

Figure 1.4: Arduino Development Environment menu [www.arduino.cc].

show the available features. The toolbar provides a wide variety of features to compose, compile, load, and execute a sketch.

1.3.5   WRITING AN ARDUINO SKETCH

The basic format of the Arduino sketch consists of a "setup" and a "loop" function. The setup function is executed once at the beginning of the program. It is used to configure pins, declare variables and constants, etc. The loop function will execute sequentially step-by-step. When the end of the loop function is reached it will automatically return to the first step of the loop function and execute again. This goes on continuously until the program is stopped.

```
//****************************************************************
void setup()
  {
  //place setup code here
  }


void loop()
  {
```

```
//main code steps are provided here
  :
  :

  }
//****************************************************************
```

Example 1.1   Let's revisit the sketch provided earlier in the chapter.

```
//****************************************************************
#define LED_PIN 13                    //name pin 13 LED_PIN

void setup()
{
pinMode(LED_PIN, OUTPUT);             //set pin to output
}

void loop()
{
digitalWrite(LED_PIN, HIGH);          //write pin to logic high
delay(500);                           //delay specified in ms
digitalWrite(LED_PIN, LOW);           //write to logic low
delay(500);                           //delay specified in ms
}
//****************************************************************
```

In the first line the #define statement links the designator "LED_PIN" to pin 13 on the Arduino processor board. In the setup function, LED_PIN is designated as an output pin. Recall the setup function is only executed once. The program then enters the loop function that is executed sequentially step-by-step and continuously repeated. In this example, the LED_PIN is first set to logic high to illuminate the LED onboard the Arduino processing board. A 500-ms delay then occurs. The LED_PIN is then set low. A 500-ms delay then occurs. The sequence then repeats.

Even the most complicated sketches follow the basic format of the setup function followed by the loop function. To aid in the development of more complicated sketches, the ADE has many built-in features that may be divided into the areas of structure, variables, and functions. The structure and variable features follow rules similar to the C programming language. The

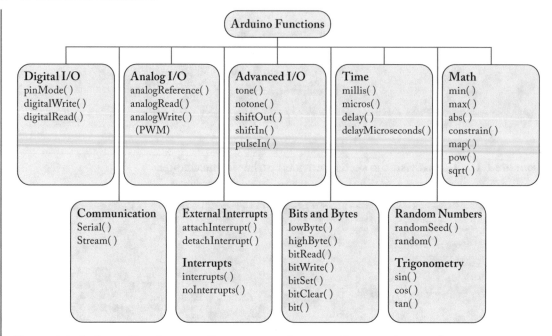

Figure 1.5: Arduino Development Environment functions [www.arduino.cc].

built-in functions consists of a set of pre-defined activities useful to the programmer. These built-in functions are summarized in Figure 1.5.

There are many program examples available to allow the user to quickly construct a sketch. These programs are summarized in Figure 1.6. Complete documentation for these programs is available at the Arduino homepage [www.arduino.cc]. This documentation is easily accessible via the Help tab on the Arduino Development Environment toolbar. This documentation will not be repeated here. With the Arduino open-source concept, users throughout the world are constantly adding new built-in features. As new features are added, they are released in future ADE versions. As an Arduino user, you too may add to this collection of useful tools. Throughout the remainder of the book we use both the ADE to program the Arduino UNO R3 and several other Arduino-based products. In the next section we get acquainted with the features of the UNO R3.

1.4   ARDUINO UNO R3 PROCESSING BOARD

The Arduino UNO R3 processing board is illustrated in Figure 1.7. Working clockwise from the left, the board is equipped with a USB connector to allow programming the processor from a host PC or laptop. The board may also be programmed using In System Programming (ISP)

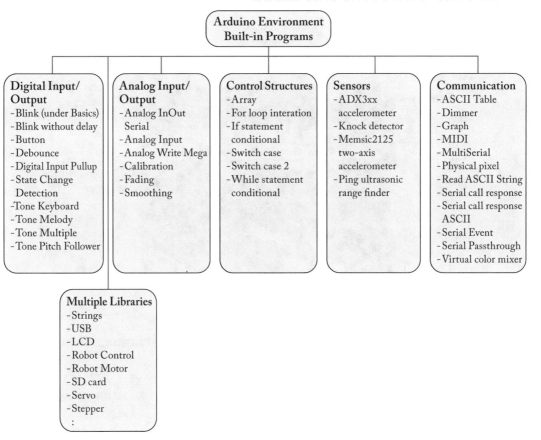

Arduino Environment Built-in Programs

Digital Input/ Output
- Blink (under Basics)
- Blink without delay
- Button
- Debounce
- Digital Input Pullup
- State Change Detection
- Tone Keyboard
- Tone Melody
- Tone Multiple
- Tone Pitch Follower

Analog Input/ Output
- Analog InOut Serial
- Analog Input
- Analog Write Mega
- Calibration
- Fading
- Smoothing

Control Structures
- Array
- For loop interation
- If statement conditional
- Switch case
- Switch case 2
- While statement conditional

Sensors
- ADX3xx accelerometer
- Knock detector
- Memsic2125 two-axis accelerometer
- Ping ultrasonic range finder

Communication
- ASCII Table
- Dimmer
- Graph
- MIDI
- MultiSerial
- Physical pixel
- Read ASCII String
- Serial call response
- Serial call response ASCII
- Serial Event
- Serial Passthrough
- Virtual color mixer

Multiple Libraries
- Strings
- USB
- LCD
- Robot Control
- Robot Motor
- SD card
- Servo
- Stepper
  :

Figure 1.6: Arduino Development Environment built-in features [www.arduino.cc].

techniques. A six-pin ISP programming connector is on the opposite side of the board from the USB connector.

The board is equipped with a USB-to-serial converter to allow compatibility between the host PC and the serial communications systems aboard the Microchip ATmega328 processor. The UNO R3 is also equipped with several small surface mount light-emitting diodes (LEDs) to indicate serial transmission (TX) and reception (RX) and an extra LED for project use. The header strip at the top of the board provides access for an analog reference signal, pulse width modulation (PWM) signals, digital input/output (I/O), and serial communications. The header strip at the bottom of the board provides analog inputs for the analog-to-digital (ADC) system and power supply terminals. Finally, the external power supply connector is provided at the bottom-left corner of the board. The top and bottom header strips conveniently mate with an Arduino shield to extend the features of the Arduino host processor.

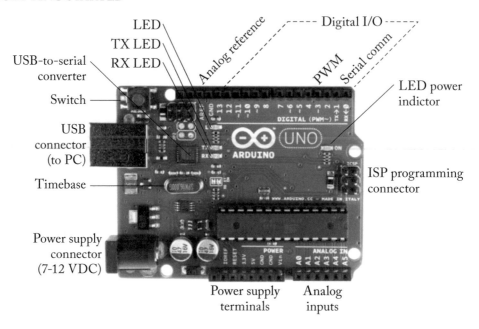

Figure 1.7: Arduino UNO R3 layout. (Figure adapted and used with permission of Arduino Team (CC BY-NC-SA) [www.arduino.cc].)

1.5   ARDUINO UNO R3 OPEN SOURCE SCHEMATIC

The entire line of Arduino products is based on the visionary concept of open-source hardware and software. That is, hardware and software developments are openly shared among users to stimulate new ideas and advance the Arduino concept. In keeping with the Arduino concept, the Arduino team openly shares the schematic of the Arduino UNO R3 processing board. See Figure 1.8.

1.6   ARDUINO UNO R3 HOST PROCESSOR–THE ATMEGA328

The host processor for the Arduino UNO R3 is the Microchip ATmega328. The "328" is a 28 pin, 8-bit microcontroller. The architecture is based on the Reduced Instruction Set Computer (RISC) concept which allows the processor to complete 20 million instructions per second (MIPS) when operating at 20 MHz. The "328" is equipped with a wide variety of features as shown in Figure 1.9. The pin out diagram and block diagram for this processor are provided in Figures 1.10 and 1.11. The features may be conveniently categorized into the following systems:

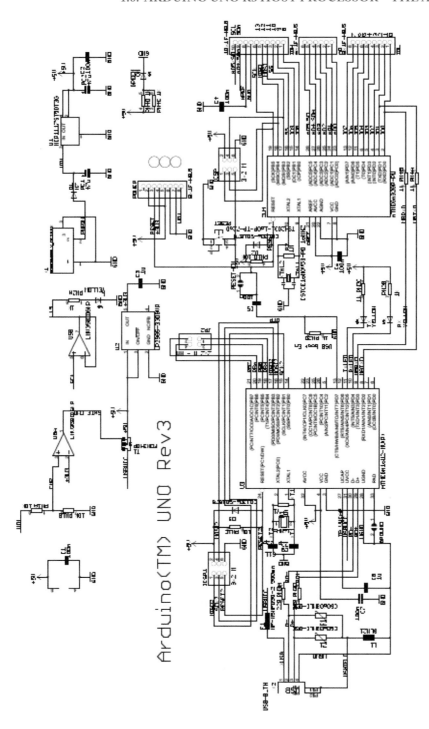

Figure 1.8: Arduino UNO R3 open source schematic. (Figure adapted and used with permission of the Arduino Team (CC BY-NC-SA) [www.arduino.cc].)

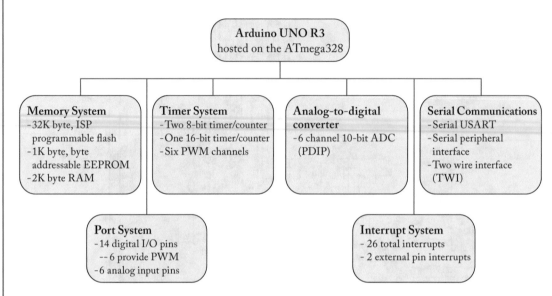

Figure 1.9: Arduino UNO R3 systems.

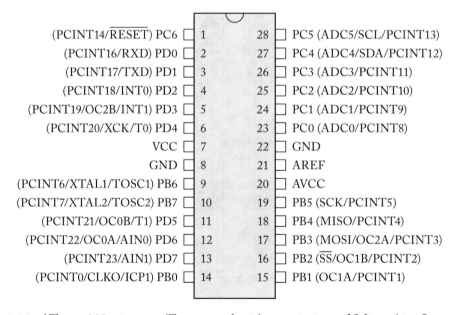

Figure 1.10: ATmega328 pin out. (Figure used with permission of Microchip, Incorporated [www.microchip.com].)

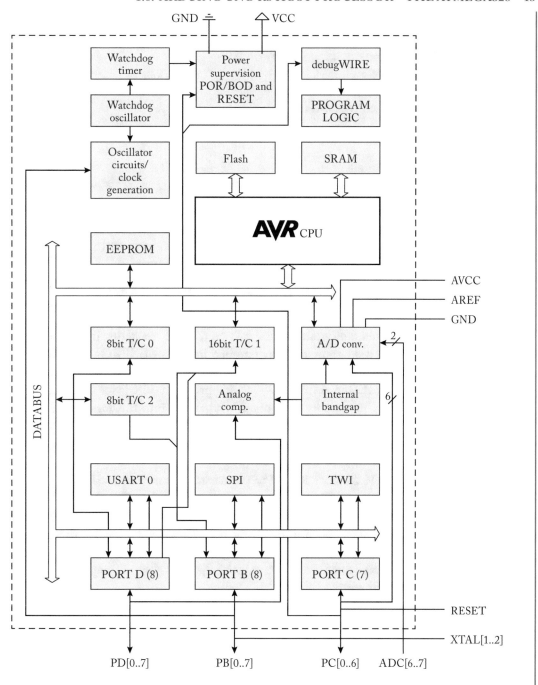

Figure 1.11: ATmega328 block diagram. (Figure used with permission of Microchip, Incorporated [www.microchip.com].)

- memory system,

- port system,

- timer system,

- analog-to-digital converter (ADC),

- interrupt system, and

- serial communications.

1.6.1   ATMEGA328 MEMORY

The ATmega328 is equipped with three main memory sections: flash electrically erasable programmable read only memory (EEPROM), static random access memory (SRAM), and byte-addressable EEPROM. We discuss each memory component in turn.

ATmega328 In – System Programmable Flash EEPROM
Bulk programmable flash EEPROM is used to store programs. It can be erased and programmed as a single unit. Also, should a program require a large table of constants, it may be included as a global variable within a program and programmed into flash EEPROM with the rest of the program. Flash EEPROM is nonvolatile meaning memory contents are retained even when microcontroller power is lost. The ATmega328 is equipped with 32K bytes of onboard reprogrammable flash memory. This memory component is organized into 16K locations with 16 bits at each location.

ATmega328 Byte – Addressable EEPROM
Byte-addressable EEPROM memory is used to permanently store and recall variables during program execution. It too is nonvolatile. It is especially useful for logging system malfunctions and fault data during program execution. It is also useful for storing data that must be retained during a power failure but might need to be changed periodically. Examples where this type of memory is used are found in applications to store system parameters, electronic lock combinations, and automatic garage door electronic unlock sequences. The ATmega328 is equipped with 1024 bytes of byte-addressable EEPROM.

ATmega328 Static Random Access Memory (SRAM)
Static RAM memory is volatile. That is, if the microcontroller loses power, the contents of SRAM memory are lost. It can be written to and read from during program execution. The ATmega328 is equipped with 2K bytes of SRAM. A small portion of the SRAM is set aside for the general-purpose registers used by the processor and also for the input/output and peripheral subsystems aboard the microcontroller. A header file provides the link between register names

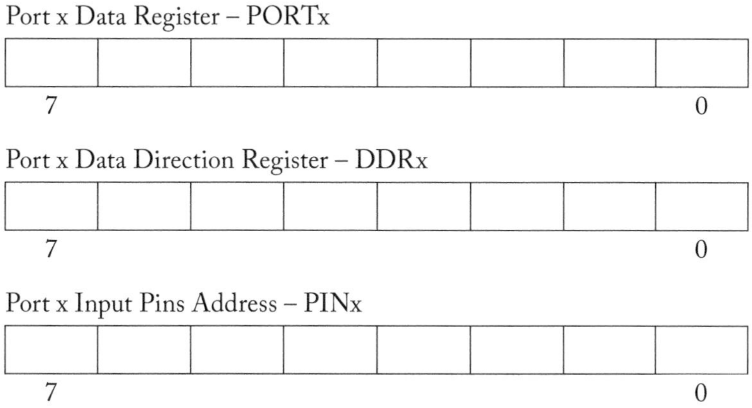

Port x Data Register – PORTx

7   0

Port x Data Direction Register – DDRx

7   0

Port x Input Pins Address – PINx

7   0

(a) Port associated registers

DDxn	PORTxn	I/O	Comment	Pullup
0	0	Input	Tri-state (Hi-Z)	No
0	1	Input	Source current if externally pulled low	Yes
1	0	Output	Output low (sink)	No
1	1	Output	Output high (source)	No

x: Port designator (B, C, D)   n: Pin designator (0–7)

(b) Port pin configuration

Figure 1.12: ATmega328 port configuration registers [www.microchip.com].

used in a program and their physical description and location in memory. During program execution, RAM is used to store global variables, support dynamic memory allocation of variables, and to provide a location for the stack.

1.6.2 ATMEGA328 PORT SYSTEM

The Microchip ATmega328 is equipped with three, 8-bit general purpose, digital input/output (I/O) ports designated PORTB (8 bits, PORTB[7:0]), PORTC (7 bits, PORTC[6:0]), and PORTD (8 bits, PORTD[7:0]). As shown in Figure 1.12, each port has three registers associated with it:

- Data Register PORTx—used to write output data to the port,

- Data Direction Register DDRx—used to set a specific port pin to either output (1) or input (0), and

- Input Pin Address PINx—used to read input data from the port.

Figure 1.12b describes the settings required to configure a specific port pin to either input or output. If selected for input, the pin may be selected for either an input pin or to operate in the high impedance (Hi-Z) mode. If selected for output, the pin may be further configured for either logic low or logic high.

Port pins are usually configured at the beginning of a program for either input or output and their initial values are then set. Usually, all eight pins for a given port are configured simultaneously.

1.6.3   ATMEGA328 INTERNAL SYSTEMS

In this section, we provide a brief overview of the internal features of the ATmega328. It should be emphasized that these features are the internal systems contained within the confines of the microcontroller chip. These built-in features allow complex and sophisticated tasks to be accomplished by the microcontroller.

ATmega328 Time Base

The microcontroller is a complex synchronous state machine. It responds to program steps in a sequential manner as dictated by a user-written program. The microcontroller sequences through a predictable fetc-decode-execute sequence. Each unique assembly language program instruction issues a series of signals to control the microcontroller hardware to accomplish instruction related operations.

The speed at which a microcontroller sequences through these actions is controlled by a precise time base called the clock. The clock source is routed throughout the microcontroller to provide a time base for all peripheral subsystems. The ATmega328 may be clocked internally using a user-selectable resistor capacitor (RC) time base or it may be clocked externally. The RC internal time base is selected using programmable fuse bits. You may choose from several different internal fixed clock operating frequencies.

To provide for a wider range of frequency selections an external time source may be used. The external time sources, in order of increasing accuracy and stability, are an external RC network, a ceramic resonator, or a crystal oscillator. The system designer chooses the time base frequency and clock source device appropriate for the application at hand. Generally speaking, if the microcontroller will be interfaced to external peripheral devices either a ceramic resonator or a crystal oscillator should be used as a time base. Both are available in a wide variety of operating frequencies. The maximum operating frequency of the ATmega328P is 20 MHz [www.microchip.com].

ATmega328 Timing Subsystem

The ATmega328 is equipped with a complement of timers which allows the user to generate a precision output signal, measure the characteristics (period, duty cycle, frequency) of an incoming digital signal, or count external events. Specifically, the ATmega328 is equipped with two 8-bit timer/counters and one 16-bit counter.

Pulse Width Modulation Channels
A pulse width modulated (PWM) signal is characterized by a fixed frequency and a varying duty cycle. Duty cycle is the percentage of time a repetitive signal is logic high during the signal period. It may be formally expressed as:

$$duty\ cycle[\%] = (on\ time/period) \times (100\%).$$

The ATmega328 is equipped with four PWM channels. The PWM channels coupled with the flexibility of dividing the time base down to different PWM subsystem clock source frequencies allows the user to generate a wide variety of PWM signals: from relatively high-frequency low-duty cycle signals to relatively low-frequency high-duty cycle signals.

PWM signals are used in a wide variety of applications including controlling the position of a servo motor and controlling the speed of a DC motor.

ATmega328 Serial Communications
The ATmega328 is equipped with a variety of different serial communication subsystems including the Universal Synchronous and Asynchronous Serial Receiver and Transmitter (USART), the serial peripheral interface (SPI), and the Two-wire Serial Interface (TWI). What these systems have in common is the serial transmission of data. In a serial communications transmission, serial data is sent a single bit at a time from transmitter to receiver. The serial communication subsystems are typically used to add and communicate with additional peripheral devices.

ATmega328 Serial USART   The serial USART may be used for full-duplex (two-way) communication between a receiver and transmitter. This is accomplished by equipping the ATmega328 with independent hardware for the transmitter and receiver. The USART is typically used for asynchronous communication. That is, there is not a common clock between the transmitter and receiver to keep them synchronized with one another. To maintain synchronization between the transmitter and receiver, framing start and stop bits are used at the beginning and end of each data byte in a transmission sequence.

The ATmega328 USART is quite flexible. It has the capability to be set to different data transmission rates known as the Baud (bits per second) rate. The USART may also be set for data bit widths of 5–9 bits with one or two stop bits. Furthermore, the ATmega328 is equipped with a hardware-generated parity bit (even or odd) and parity check hardware at the receiver. A single parity bit allows for the detection of a single bit error within a byte of data. The USART may also be configured to operate in a synchronous mode.

ATmega328 Serial Peripheral Interface (SPI)   The ATmega328 Serial Peripheral Interface (SPI) can also be used for two-way serial communication between a transmitter and a receiver.

In the SPI system, the transmitter and receiver share a common clock source. This requires an additional clock line between the transmitter and receiver but allows for higher data transmission rates as compared to the USART.

The SPI may be viewed as a synchronous 16-bit shift register with an 8-bit half residing in the transmitter and the other 8-bit half residing in the receiver. The transmitter is designated the master since it is providing the synchronizing clock source between the transmitter and the receiver. The receiver is designated as the slave.

ATmega328 Two-wire Serial Interface (TWI)   The TWI subsystem allows the system designer to network related devices (microcontrollers, transducers, displays, memory storage, etc.) together into a system using a two-wire interconnecting scheme. The TWI allows a maximum of 128 devices to be interconnected. Each device has its own unique address and may both transmit and receive over the two-wire bus at frequencies up to 400 kHz. This allows the device to freely exchange information with other devices in the network within a small area.

ATmega328 Analog to Digital Converter (ADC)

The ATmega328 is equipped with an eight-channel analog to digital converter (ADC) subsystem. The ADC converts an analog signal from the outside world into a binary representation suitable for use by the microcontroller. The ATmega328 ADC has 10-bit resolution. This means that an analog voltage between 0 and 5 V will be encoded into one of 1024 binary representations between $(000)_{16}$ and $(3FF)_{16}$. This provides the ATmega328 with a voltage resolution of approximately 4.88 mV.

ATmega328 Interrupts

The normal execution of a program follows a specific sequence of instructions. However, sometimes this normal sequence of events must be interrupted to respond to high priority faults and status both inside and outside the microcontroller. When these higher priority events occur, the microcontroller suspends normal operation and executes event specific actions contained within an interrupt service routine (ISR). Once the higher priority event has been serviced by the ISR, the microcontroller returns and continues processing the normal program.

The ATmega328 is equipped with a complement of 26 interrupt sources. Two of the interrupts are provided for external interrupt sources while the remaining interrupts support the efficient operation of peripheral subsystems aboard the microcontroller.

1.7   INTERFACING TO OTHER DEVICES

Chapter 3 "Arduino Power and Interfacing" of *Arduino I: Getting Started* introduces the extremely important concept of microcontroller interfacing. Anytime an input or an output device is connected to a microcontroller, the interface between the device and the microcontroller must be carefully analyzed and designed. This will ensure the microcontroller will continue to oper-

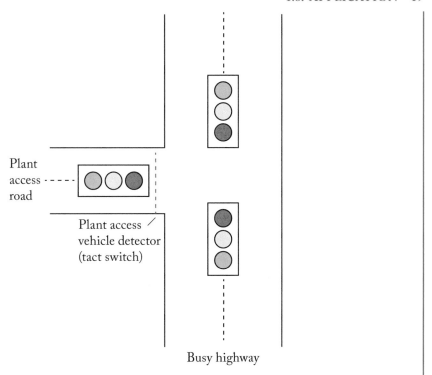

Figure 1.13: Intersection.

ate within specified parameters. Should the microcontroller be operated outside its operational envelope; erratic, unpredictable, and an unreliable system may result.

1.8   APPLICATION

Inexpensive hardware may be used as a first prototype of more complex systems. In this exercise we use LEDs and switches to develop a first prototype of a traffic control system.

Scenario. A busy two-lane highway exits from the south side of a small community. Near the city's edge is a manufacturing plant that runs three 8 hour shifts. The entrance to the plant is via a small road intersecting the busy highway. Develop and test an algorithm to control a traffic light at the intersection of the busy highway and plant access road. Normally, the busy highway will have priority for through traffic. However, at shift changeover time windows, vehicles exiting the plant will be allowed access to the highway.

Model the intersection light using LEDs. A vehicle ready for exit on the plant access road is indicated using a tact momentary switch, as shown in Figure 1.13.

- **Part 1.** Develop an Arduino sketch to control the traffic lights at the intersection.

- **Part 2.** Develop a control algorithm for the intersection of two, 2-lane busy streets within a city. Provide for pedestrian crosswalks.

1.9   SUMMARY

The goal of this chapter was to provide an introduction and tutorial on the Arduino UNO R3. We used a top-down design approach. We began with the "big picture" of the chapter followed by an overview of the ADE and the subsystems aboard the ATmega328.

1.10   REFERENCES

[1] Arduino homepage. www.arduino.cc

[2] *Microchip ATmega328 PB AVR Microcontroller with Core Independent Peripherals and Pico Power Technology DS40001906C*, Microchip Technology Incorporation, 2018. www.microchip.com

1.11   CHAPTER PROBLEMS

1.1. Describe the steps in writing a sketch and executing it on an Arduino UNO R3 processing board.

1.2. What is the serial monitor feature used for in the Arduino Development Environment?

1.3. Describe what variables are required and returned and the basic function of the following built-in Arduino functions: Blink, Analog Input.

1.4. Sketch a block diagram of the ATmega328 and its associated systems. Describe the function of each system.

1.5. Describe the different types of memory components within the ATmega328. Describe applications for each memory type.

1.6. Describe the three different register types associated with each port.

1.7. How may the features of the Arduino UNO R3 be extended?

1.8. Discuss different options for the ATmega328 time base. What are the advantages and disadvantages of each type? Construct a summary table.

1.9. Discuss the three types of serial communication systems aboard the ATmega328. Research an application for each system.

1.10. What is the difference between an ADC and DAC. Which one is aboard the ATmega328?

1.11. How does the Arduino UNO R3 receive power? Describe in detail.

1.12. What is the time base for the Arduino UNO R3? At what frequency does it operate? How many clock pulses per second does the time base provide? What is the time between pulses?

1.13. What is meant by the term open source?

1.14. What is the maximum operating frequency of the ATmega328? What is the lowest operating frequency of the ATmega328?

1.15. What is the range of operating voltages that may be applied to the ATmega328?

CHAPTER 2

The Internet and IoT

Objectives: After reading this chapter, the reader should be able to do the following.

- Describe how data is routed and exchanged via the internet.

- Summarize significant events in internet development.

- Describe protocol stacks used in internet message processing.

- Define different methods of device addressing employed within the internet.

- Define the function of key internet hardware components.

- Summarize the differences between network configurations.

- Summarize threats to safe and secure internet operations.

- Provide a definition for the Internet of Things (IoT) concept.

- Provide an IoT feature list.

- Define Information Technology (IT) and Operational Technology (OT).

- Describe the features of a programmable logic controller (PLC) based OT system.

- Provide a working definition of the Industrial Internet of Things (IIoT).

- Describe how IoT concepts are employed within a specific industry.

- Sketch an IoT model and describe the relationship between "things" and applications.

2.1 OVERVIEW

Welcome to the world of the internet! This chapter provides a condensed overview of internet operation and history. We begin with a big picture overview of the internet followed by a brief review of internet development. We then take a closer look at selected internet topics including protocols, addressing techniques, different network configurations, threats to internet security, and threat countermeasures. Admittedly, the treatment of these topics is brief. We discuss internet concepts related to our exploration of the IoT. We then discuss IoT concept applications in the industrial world of the IIoT. For additional details, the interested reader is referred to the excellent references at the end of the chapter used to compile the information presented here.

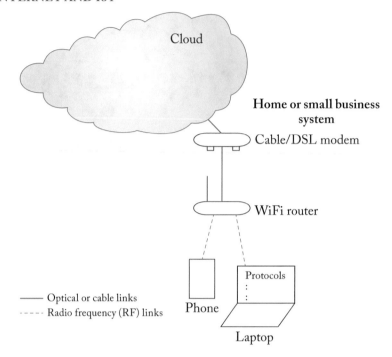

Figure 2.1: Internet at home or a small business configuration.

2.2   A BIG PICTURE OF THE INTERNET

From its early beginnings in the late 1960s to today, the internet has become ubiquitous (found everywhere) in every facet of our lives. A few examples where the internet has become prevalent include industry, agriculture, energy production, education, healthcare, entertainment, manufacturing, retail, communications, and many other areas (Hanes [5]).

Many (including the author) take a safe, secure, and reliable internet for granted. Portions of the internet, consisting of a global network of interconnected computers, are referred to as the "cloud." In this section we examine connections between computers that comprise the internet in a home and work environment and then examine what is inside the cloud.

Figure 2.1 provides a typical internet connection found in a home or small business such as a cafe or small store. A cable or digital subscriber line (DSL) modulator/demodulator (modem) provides a connection to the internet. The cable/DSL modem is provided by an Internet Service Provider (ISP) when you subscribe to their internet connection service. The connection between the cable/DSL modem and the ISP provider may be a combination of copper cable, optical fiber, and wireless radio frequency connection links.

With internet service available via the cable/DSL modem, a WiFi router is used to establish a wireless local area network (WLAN) within your home or small business. The WLAN

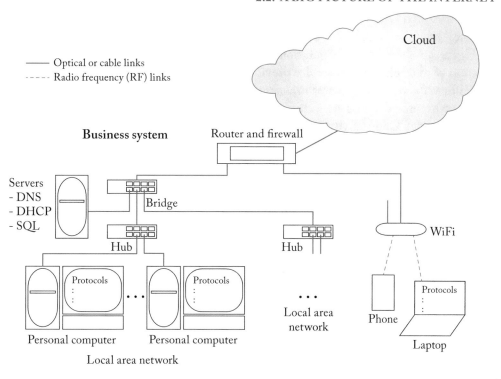

Figure 2.2: Internet at a large business or university.

serves as an internet access point to a broader area using radio frequency (RF) signals operating at 2.5 or 5 GHz. The link between the cable/DSL modem to the WiFi router is via a cable. Wireless devices such as a cell phone or a laptop within range of the WiFi router are able to access the internet with proper credentials (i.e., WLAN password and ISP subscription). The WiFi router range may be extended using a WiFi range extender.

Internet service typically found in a larger business or a university is shown in Figure 2.2. Internet service is provided to the organization via an ISP. Connection is made to the ISP via a firewall. The firewall provides protection from internet hazards outside the organization. It may also be used to limit outgoing information from the organization (e.g., sensitive company information, classified material, etc.). There is also a router at the organization portal. It is used to route internet message traffic to its next destination.

Individual computers are provided access to the internet via a cabled connection to a hub. The hub is used to connect computers into a local area network (LAN) sharing the same location or similar function such as an academic department or company section. Multiple LANs are then connected via a bridge. The bridge will also provide connection to a series of local servers such as a Domain Name System (DNS) server, a Dynamic Host Configuration Protocol (DHCP) server, a Structured Query Language (SQL) database server, and a mail server (among others). WiFi access may also be provided by a WiFi router as previously described.

Figure 2.3 shows the configuration of the internet cloud. The cloud contains the global connection of multiple internet service providers. Regional internet service providers share internet traffic via a metropolitan area exchange (MAE). The regional network ISPs connect to a network service provider who are in turn connected to other network service providers via network access points. Connectivity across the globe is provided by submarine optical fiber cables spanning the oceans. Internet access may be provided to remote areas via balloon borne network access points (Loon [11]).[1] The overall result is a global network of interconnected computers for the open exchange of information.

Aside from the internet hardware components, there are internet protocols and applications used to insure reliable and compatible communications from one location to another. As an example, when you are checking your favorite news website or sending an e-mail to a friend, the specific application you are using along with the computer's operating system has built in features to interact with the internet to accomplish your desired task.

The application and the operating system apply the different layer activities to transmit information over the internet. For example, the e-mail message you are sending is broken into packets of information, provided source, and destination IP addresses, and converted to an electronic signal. The packets are then routed to the destination computer via a series of internet hops directed by routers along the path. At the destination the received packets are reassembled and the protocol steps are applied in reverse order.

2.3   BRIEF HISTORY

The internet started as a United States government project in the late 1960s. The Advanced Research Projects Agency (ARPA) sponsored a project to interconnect four computers at different locations to share resources as shown in Figure 2.4. The resulting network was dubbed ARPANET.

From this early start additional computers were added and evolved into the Defense Advanced Research Projects Network or DARPANET. In the mid-1980s the National Science Foundation established NSFNET. The early government sponsored programs were the first Internet Service Providers. Commercial providers joined in the late 1980s with full commercial control of the internet by 1995. The Internet Society (ISOC) was established in 1992 to assure all people globally benefit from the internet (Levine [10]).

From this basic understanding of internet operation and history, we spend the remainder of the chapter taking a more detailed view of selected topics necessary for our investigation of the IoT.

---

[1]January 2021 news releases indicate the Loon system may be phased out.

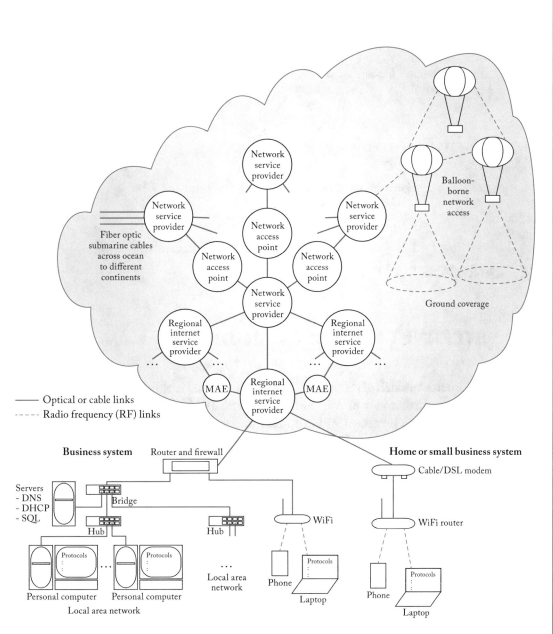

Figure 2.3: The cloud.

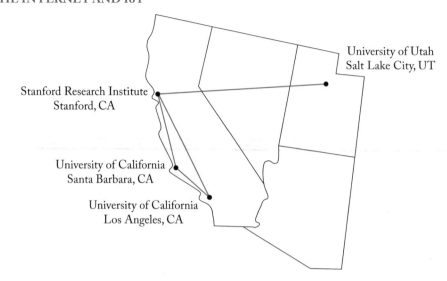

University of Utah
Salt Lake City, UT

Stanford Research Institute
Stanford, CA

University of California
Santa Barbara, CA

University of California
Los Angeles, CA

Figure 2.4: ARPANET.

2.4   INTERNET PROTOCOL MODELS

There are two different models of protocol stacks commonly used within the internet community. A protocol is a standardized set of rules and procedures. The layered protocol models provide guidelines on how data is processed within applications and prepared for transmission over the internet. Both protocols were developed in the early 1980s.

The International Organization for Standardization (ISO) developed the seven layered Open Systems Internet protocol stack ISO/OSI reference model shown in Figure 2.5a. The adjacent layers interact with one another in a given system while similar layers interact with one another in different systems.

The transmission control protocol/internet (TCP/IP) maps into several layers of the OSI model as shown in Figure 2.5a. Figure 2.5b shows the four layers of the TCP/IP model: application, transport, internet, and link layer.

The data to be shared with another computer resides within the application layer. The data is divided into packets. As the data packet is processed through each layer of the sending computer, additional header and footer information is appended to the data payload. The IP address allows the sender and receiver to find one another on the internet (Leiden [9], Lowe [12], and Null [14]).

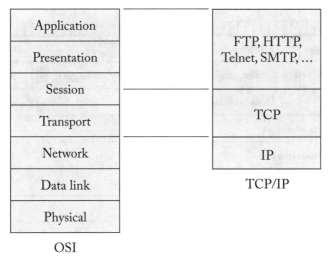

(a) Comparison of the OSI protocol stack model to the TCP/IP protocol stack model

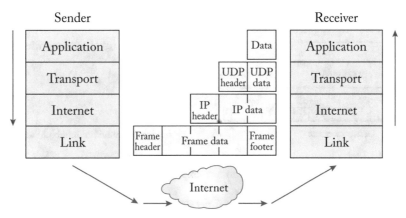

(b) TCP/IP protocol for sender and receiver interaction

Figure 2.5: OSI vs. TCP/IP (Leiden [9], Lowe [12], and Null [14]).

2.5   INTERNET ADDRESSING TECHNIQUES

In this section we discuss the importance of and techniques used to address network assets. We begin with IP addressing and packet headers. There are two different versions of the IP header: IPv4 and IPv6, as shown in Figure 2.6.

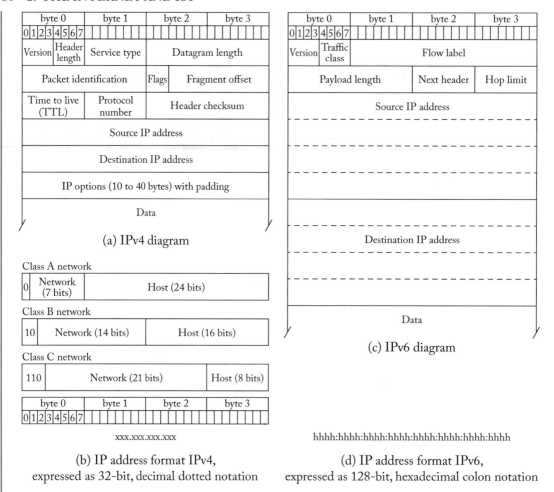

(a) IPv4 diagram

(c) IPv6 diagram

(b) IP address format IPv4,
expressed as 32-bit, decimal dotted notation

(d) IP address format IPv6,
expressed as 128-bit, hexadecimal colon notation

Figure 2.6: IPv4 and IPv6 (Leiden [9] and Null [14]).

2.5.1   IPv4 HEADER

The earlier IPv4 header version consists of 24 bytes followed by the data payload. The overall packet datagram must be at least 40 bytes. The IPv4 header consists of the following fields:

- Version: IP protocol version. For IPv4 this field is set to $(0100)_2$.

- Header length: specified as 32-bit words.

- Type of service: specifies priority from low (000) to critical (101).

- Total length: specifies total length of the datagram packet in bytes.

- Packet ID: the packet is assigned a unique serial number.

- Flags: indicates whether a larger packet can be broken into smaller packets.

- Fragment offset: provides fragment location within packet.

- Time to live: determines number of internet hops are allowed from source to destination.

- Protocol number: indicates the type of protocol associated with the data: 0–reserved, 1–internet control message protocol (ICMP), 6–transmission control protocol (TCP), or 17–user datagram protocol (UDP).

- Header checksum: holds the calculated checksum of the header.

- Provides source and destination address of packet. Note each address is 32 bits in length. A unique address or source ID is provided to each source computer on the internet. The allocation of addresses is coordinated by the Internet Corporation for Assigned Names and Numbers or ICANN [www.ICANN.org].

- IP options: provides additional control information.

The IPv4 protocol uses the 32-bit IP address configuration as shown in Figure 2.6b. The first several bits of the address indicates the network class: Class A for large networks (0), Class B for medium-sized networks (10), and Class C for smaller networks (110). The remaining IP address bits are partitioned to select a network and a specific host. The IP address is expressed as a 32-bit dotted decimal notation value (xxx.xxx.xxx.xxx). Each byte in the address is specified by its decimal equivalent (xxx) and has a value ranging from 0–255 (Hanes [5], Lowe [12], and Null [14]).

2.5.2   CIDR ADDRESSING

To provide additional addressing flexibility a Classless Inter-Domain Routing (CIDR) addressing scheme was developed. The CIDR scheme provides for flexible subnet addressing within the IPv4 protocol. The format of the CIDR address is provided in Figure 2.7a. A subnet is a smaller network within one of the network classes: A, B, or C.

Recall the IPv4 protocol provides for a 32-bit IP address. The CIDR protocol allows the address to be partitioned into a number of address bits allocated to identify the network and the remaining bits allocated to identify a specific host (computer) within the network. A slash character (/) follows the address with a decimal number. The decimal number indicates the number of logic ones in the subnet mask.

In the example provided in Figure 2.7b left an IP address has been partitioned such that 24 bits will be used to identify the specific network while the remaining 8 bits will be used to specify a specific computer within the network. This partition allocation results in a subnet mask of

Number = leading 1's in routing mask

(a) Classless inter-domain routing (CIDR) addressing format

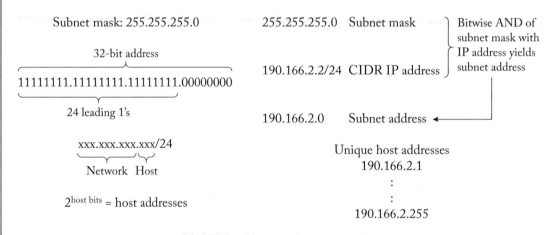

(b) CIDR addressing format example

Figure 2.7: CIDR addressing [www.IETF.org].

255.255.255.0 or expressed in separated binary as $(11111111.11111111.11111111.00000000)_2$. Since there are 24 bits in the subnet mask containing leading ones, a /24 is appended to the IP address to communicate the partition information. The 8 bits allocated for host or computer addressing will provide for $2^{host\ bits}$ or 256 unique addresses.

In Figure 2.7b right, the IP address (190.166.2.2./24) has been partitioned into 24 bits for the subnet address and the other 8 bits for specific computer addressing. This partition results in the subnet mask 255.255.255.0. When this mask is logically ANDed with the IP address the resulting subnet address 190.166.2.0 results. The specific computers within the subnet will be addressed beginning at 190.166.2.1. and ending at 190.166.2.255 [www.IETF.org].

2.5.3   IPv6 HEADER

In the mid-1990s, the IPv6 was released by the Internet Engineering Task Force (IETF). The IPv6 protocol provides for a longer 128-bit IP address space. The different fields within the IPv6 header specify the following.

- Version: IP protocol version. For IPv6 this field is set to $(0110)_2$.

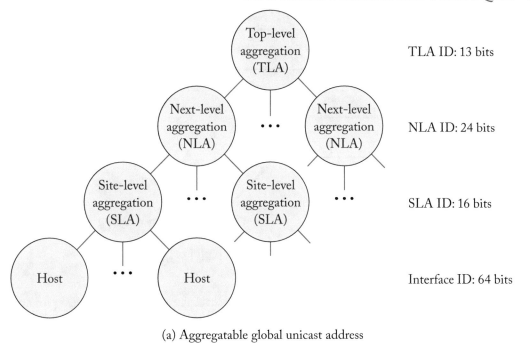

(a) Aggregatable global unicast address

Prefix 001 3 bits	TLA ID 13 bits	Reserved 8 bits	NLA ID 24 bits	SLA ID 16 bits	Interface ID 64 bits

(b) Aggregatable global unicast address format

Figure 2.8: IPv6 addressing (Hanes [5], Lowe [12], and Null [14]).

- Traffic class: will specify different priority.

- Flow label: will specify the type of communication in progress.

- Payload length: expressed in bytes

- Next header: specifies if additional header information is provided in the payload.

- Hop limit: will allow up to 256 hops from source to destination.

- Source and destination addresses: 128 bits each.

The IETF developed a logical, methodical method of assigning IPv6 addresses, the Aggregatable Global Unicast Address Format, as shown in Figure 2.8. The 128-bit address is parti-

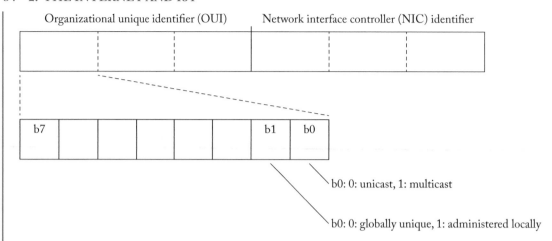

Figure 2.9: MAC address (Hanes [5], Lowe [12], and Null [14]).

tioned into different fields to specify top, next, and site-level aggregation representing for an example a country, a company within the country, and networks within the company, respectively. The 64-bit interface ID is a combination of the host device MAC address and information from a nearby router. The 128-bit IPv6 address is specified as eight 16-bit values expressed in hexadecimal and separated by colons (hhhh:hhhh:hhhh:hhhh: hhhh:hhhh:hhhh:hhhh) (Hanes [5], Lowe [12], and Null [14]).

2.5.4   MAC ADDRESS

The host device's Medium Access Control (MAC) address is a 48-bit device specific address as shown in Figure 2.9. The address is partitioned into six different bytes. The address specifies the Organizational Unique Identifier (OUI) and the Network Interface Controller (NIC) identifier. The NIC provides the interface between the host computer or device and the internet. The MAC addressing scheme allows each NIC to have a unique address (Hanes [5], Lowe [12], and Null [14]).

2.5.5   DNS AND URL ADDRESSING

Rather than memorize the IP address for individual networks and computers, descriptive, user-friendly names may be assigned using Domain Name System (DNS) techniques. The DNS serves as a distributed directory of named network assets. The directory is stored on a number of DNS servers throughout the internet. You may apply for a DNS name from a DNS provider. The provider will determine if the name is available for use. The Internet Corporation for Assigned Names and Numbers (ICANN) coordinates the use of DNS names across the globe (Shuler [16]).

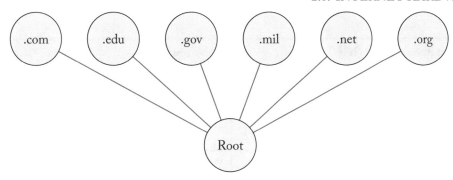

Figure 2.10: Common URL domains (Shuler [16]).

To completely specify the location of a computer on the internet, a Uniform Resource Locator (URL) address is used. The URL address consists of three parts: the protocol identifier, the DNS name, and the domain name.

A common protocol is "http." The protocol type is followed by ": $//www..$" The next portion of the URL address is the DNS name followed by the domain (e.g., .edu, .org, etc.). URL domains are shown in Figure 2.10. As an example, the website address for the main Arduino site is: http://www.arduino.cc. The ".cc" is a variant of the ".com." domain.

2.6   INTERNET HARDWARE

We began the chapter with a brief overview of internet operation. In this section we provide a brief summary of hardware components used to make the internet operate correctly. Key internet components are summarized in Figure 2.11. For each internet component we provide the Cisco Systems icon to sketch an internet layout diagram.

2.7   CYBERSECURITY

One of my favorite books is the *The Once and Future King* by T. H. White. It is the tale of the young boy, "the Wart," becoming King Arthur and the many adventures along the way. I have read this book every several years since I was young. Early in the book, White provides a description of the Wart's guardian's castle. He describes how the castle is protected from marauders by a moat (deep ditch) filled with water. To get access to the castle, a drawbridge is lowered across the moat and then raised again to secure the castle.[2]

There are many dangers surrounding a network or a computer on the internet. As shown in Figure 2.12, the dangers are in the form of malicious software (malware) or the nefarious efforts of computer hackers. These dangers and challenges include (Kurose [8], Levine [10], and Lowe [12]):

---

[2]T. H. White, *The Once and Future King*.

Network interface card (NIC)	Cable modem
Bridges physical network and computer Connected to computer I/O bus Converts to network compatible signal - parallel to serial conversion Contains unique MAC address	Internet entry point to Internet Service Provider (ISP) Linked to ISP via coaxial cable May be equipped with router features
Repeater	**WiFi router**
Correct signal attenuation - reconstitute digital signal - amplifies signal	Provides WiFi access to ISP for homes and small businesses Connects to cable modem
Hub	**Storage server**
Multiple inputs and outputs Process individual packets - receives input packet - broadcasts packets network device(s)	Provide storage service for network
Switch	**Directory server**
Connects selected input and output port Handles multiple communication links simultaneously	Directory of DNS name to IP address
Bridge	**Router**
Links two similar network segments Result: same subnet, IP address prefix Each node must have unique MAC address Isolates collision domains	Connects networks of different media Determines packet route to destination Uses O/S routing tables to direct next hop Exchanges traffic between networks Serves as boundary between subnets
Gateway	**Firewall**

Network point of entry
Provides communication services at
   all seven OSI layers
Converts protocols, character codes
Performs encryption/decryption
	Protects internet from external dangers
	Applied policy to determine allowable
	   trafffic
	Controls message traffic in and out of
	   network

Network icons used courtesy of CISCO Systems

Figure 2.11: Internet hardware. (Cisco Systems icons used courtesy of Cisco Systems, Inc.)

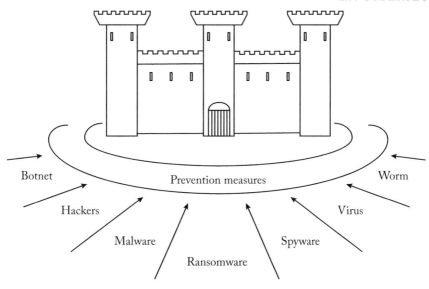

Figure 2.12: Network threats.

- botnet—network of infected computers controlled from an external source to perform coordinated nefarious activities on target computers;

- hackers—individuals who try to overcome computer protection measures and procedures to gain personal data;

- ransomware—a computer attack where files are encrypted and held for ransom. If the ransom is paid, the files are returned to normal service;

- spyware—software that is accidently downloaded while browsing the internet. The software spies on your computer activities and reports back to its source;

- virus—a nefarious software program spread as an e-mail attachment. When the e-mail attachment is executed it goes to your computer's address bank and sends out e-mails with the virus program as an attachment masquerading as you. Using this technique, the virus may be spread to a number of computers. The nefarious intent of the virus may be activated by a specific event such as reaching a particular date and time; and

- worm—a worm creeps into a computer by means of flaws within network programs. Once onboard your computer, the worm looks for password and credit card information.

As in the analogy, the castle is protected from marauders by the surrounding moat and securing the drawbridge. The moat and drawbridge for a network and its computer assets include preventive countermeasures including the following (Kurose [8], Levine [10], and Lowe [12]).

- Firewall—A firewall protects network resources from external dangers. It applies policies to determine message traffic that may enter a protected network.

- Antivirus programs (AVP)—Each computer on the network should have an AVP installed. The AVP should be current with all software updates applied.

- Operating system updates—Regular operating system updates are sent to computer users. These updates should be made when received. They may contain updates to correct a security flaw.

- Passwords—You should employ a strong password to protect your computer assets. IoT hardware devices are sometimes configured with a default password. You should replace the default password with a strong password.

- File backups—Computer files should be backed up on a regular basis.

- User awareness—Users should be skeptical of e-mails that appear questionable. An e-mail with an executable attachment should not be opened.

With this brief introduction to the internet, we now examine how internet infrastructure is employed to support the IoT and IIoT.

2.8   INTERNET OF THINGS (IoT)

In this section we provide a brief introduction to the fascinating world of the IoT. We begin with a study of an IoT definition from leading experts. We then attempt to distinguish between the concept of IoT and Cyber-Physical Systems (CPS). Next, we investigate the difference and overlap between the two main branches of IoT: Information Technology (IT) and Operational Technology (OT). To provide order to all the new concepts, a simplified IoT architecture is discussed. The pervasiveness of IoT is then examined in industry or the IIoT.

The term Internet of Things was first used by Kevin Ashton in a 1999 Proctor & Gamble presentation. Mr. Ashton's presentation discussed concepts on using the existing internet infrastructure to support P&G's supply chain (Greer [4] and Hanes [5]). From this early start, applications within business and industry have become quite pervasive.

A review of the literature provides a feature list describing the IoT systems concept (Rajkumar [15], Hanes [5], and Greer [4]):

- an IoT system connects things to the internet;

- each thing or device has its own unique identifier or address;

- communication between things is provided via the internet;

- an IoT system provides for interrelated and integrated computing devices and physical processes;

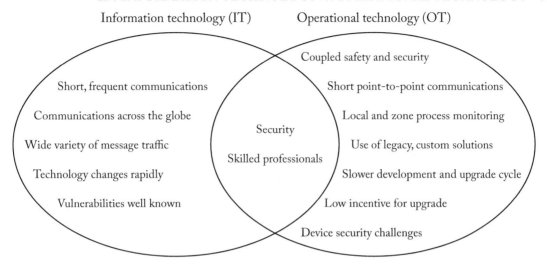

Figure 2.13: Information Technology (IT) vs. Operational Technology (OT) (Hanes [5]).

- an IoT system provides the ability to measure, process, and transfer to and from remote locations; and

- IoT processes are monitored, coordinated, and controlled.

Interestingly, the concept of CPS share many of the same features. The National Institute of Standards and Technology (NIST) performed a study to examine the relationship between IoT and CPS and noted although the concepts originated in different industries, they are substantially equivalent concepts. A unified perspective of the two concepts was provided: "Internet of Things and Cyber-Physical Systems comprise interacting logical, physical, transducer, and human components engineered for function through integrated logic and physics" (Greer [4]).

As an end of chapter exercise, we ask you to develop your own working definition of the IoT.

2.9 INFORMATION TECHNOLOGY vs. OPERATIONAL TECHNOLOGY

A key concept within IoT is the close relationship between IT and OT. Earlier in the chapter a brief introduction to the internet and the related concepts within IT was provided. A related but distinct field is OT. The relationship between IT and OT are shown in Figure 2.13.

As discussed earlier in the chapter, IT communications usually consist of short, frequent communications that are broken into packets and communicated globally. IT provides a wide variety of message traffic including e-mails, requests, and response for information from websites, and multiple other types. IT technology developments are rapidly evolving with vulnerabilities well known, documented, combatted, and corrected (Hanes [5]).

OT provides for process control within many areas of industry. As shown in Figure 2.13, industrial safety and security are intertwined. OT communications are typically short, point-to-point communications on a factory floor or within an industrial process. Monitoring via a Supervisory Control and a Data Acquisition (SCADA) system is typically performed within a local and confined zone. Although OT developments are actively taking place, adoption time-lines are slower that with IT. Since OT governs a proprietary and custom solution for a given industrial process, typically there is a low incentive for technology upgrade. Although many IT and OT concepts are related but different, IT and OT both enjoy the dedication of skilled professional practitioners (Hanes [5]).

IT and OT share the requirement for robust security protection and countermeasures. Many of the security concepts discussed for IT also apply for OT. In the industrial world, IT and OT systems are often linked to share information among related processes. For example, a remote oil drilling platform may be controlled via OT processes. If an oil company has multiple remote platforms, they may be linked via IT processes to share production data. Some form of isolation, an "air gap," is typically provided between IT and OT related processes for security purpose. This helps prevent a nefarious actor from accessing a critical industrial process via the internet (Hanes [5]).

2.10   OPERATIONAL TECHNOLOGY

OT is used to control industrial processes. The fundamental OT building block is the programmable logic controller (PLC). A PLC diagram is provided in Figure 2.14a.

A PLC is an industrial hardened microcontroller. As shown in Figure 2.14a, a PLC is typically a rack-mounted collection of modules. Each module provides a critical subsystem for the PLC. The PLC subsystems share many of the same functions typically found in most microcontrollers. For example, a typical PLC system consists of power supply, CPU, serial communications, analog input, digital input and output, and timer modules. A custom system is assembled by choosing modules to meet system requirements.

PLC systems are typically programmed using ladder logic techniques. A ladder logic program resembles a ladder with two vertical side rails linked by a number of rungs. As shown in Figure 2.14b, the real-world inputs (switches, sensors, etc.) are interfaced to the PLC via input modules. Output real-world devices such as indicators, audible alarms, motors, actuators, etc., are interfaced to the PLC via output modules. The PLC ladder logic rungs represent steps of a program that link input device conditions to output control signals.

As shown in Figure 2.14c, a ladder logic program goes through a scan consisting of multiple stages. The scan begins with an input scan. During the input scan, the status of inputs is checked and an input image table is updated in the PLC CPU memory. The input status is fixed in the input image table for the remainder of the scan time. With input status updated, the

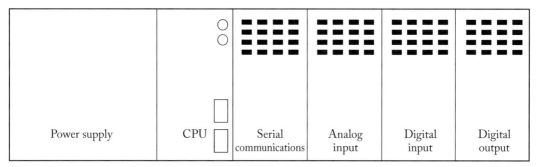

(a) Modular programmable logic controller system

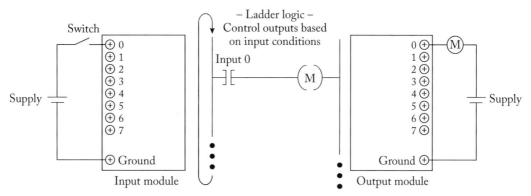

(b) Ladder logic links control outputs based on the status of input conditions and the linking PLC instructions [Stenerson]

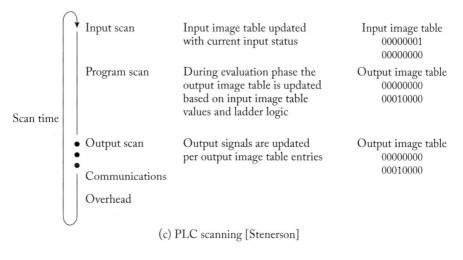

(c) PLC scanning [Stenerson]

Figure 2.14: **PLC overview** (Stenerson [17]).

program scan commences. This is called the evaluation phase where the output image table is updated based on the input image table values and the ladder logic rungs connecting input values to output control signals. Each rung in the ladder logic program is evaluated sequentially starting with the top rung and progressing down the ladder. With the completion of the evaluation stage, output signals are generated per the output image table. The final two steps of the scan include related serial communications and any required PLC housekeeping. Upon completion of the scan, the scan is repeated beginning again with the input scan (Stenerson [17]).

2.11   IoT ARCHITECTURE

The IoT, as first described by Ashton in 1999, initiated the movement to provide a link between the IT and OT worlds. There are multiple models available to describe this vital link. Hanes et al. [5] provides the model shown in Figure 2.15a.

The model provides three layers linking IoT "things" to applications via internet-based communication channels. The "things" are the sensors and actuators interfacing to a physical world process. The sensors and actuators provide for the monitoring and control signals for the process. The application, which may be physically distant from the process, takes in as input the sensor information and provides output control signals based on the control algorithm (Hanes [5]).

Arduino provides a similar model as shown in Figure 2.15b. In this chapter's Application section we follow the steps on the model's left side to: register a device, connect the device to the IoT cloud, add device properties, and edit and deploy the sketch. Once deployed, the sketch interacts with the physical world to collect data via sensors. The data collected are sent to the cloud-based application as an event. Based on the data collected, the application algorithm generates control signals. The signals are then sent as an event back to the physical process.

Example.   I have always wanted to build a greenhouse. I find it quite fascinating that the sun's energy may be captured, stored, and employed at a later time to extend and stabilize the growing season for vegetables. Part of the fascination may be related to spending much of my life in northern climes (Newfoundland, Nebraska, North Dakota, Montana, and Wyoming).

Applying the IoT models described, the "things" of the greenhouse would be the sensors used to measure the vital signs of the greenhouse. For example, we might measure the following parameters: indoor temperature, outdoor temperature, humidity, soil moisture, stored water level, backup battery voltage level, etc. The actuator "things" of the greenhouse would be those devices used to change the greenhouse configuration: a vent fan when the indoor greenhouse temperature becomes too high, a water pump to mist the vegetables when appropriate conditions are met (e.g., plant soil too dry, etc.). An Arduino-based sketch may be developed to visualize and manage greenhouse properties. For example, the greenhouse indoor and outdoor temperatures may be logged and displayed over a long period of time (e.g., the winter months).

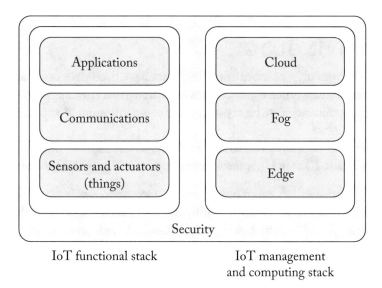

(a) Simplified IoT model [Hanes]

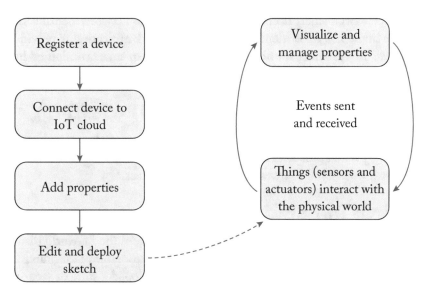

(b) Arduino IoT deployment model [arduino.cc]

Figure 2.15: **IoT models** (Hanes [5], Arduino.cc).

The internet infrastructure with WiFi access may be used to allow the sending and receiving of greenhouse events. In Chapter 4, we explore this IoT application in great detail.

2.12   IoT TECHNOLOGY

To support IoT deployment, a number of technologies have been developed to support project level IoT applications, smart home concepts, and industrial-level IIoT applications. The dividing lines between these applications are blurry. It is more of a continuum of applications rather than categories of applications.

Later in the chapter, we get acquainted with the MKR 1010 in the Application section. We also employ a MKR 1000 in Chapter 4 to develop an IoT greenhouse control system.

Smart Home Applications.   A smart home uses technology to efficiently monitor and control home parameters such as temperature, humidity, lighting, security, lawn health, etc. In 2005, the Z-Wave Alliance was established to provide a standard configuration and control protocol for smart home applications. The Z-wave protocol provides for the wireless mesh networking of smart objects within a home. The protocol provides for a data rate between configured devices of 100 kbps. Devices communicate securely at frequencies of 908.4 or 916 MHz using AES 128 encryption.[3] Network activities are coordinated by a smart hub that is connected to the internet. The smart hub can control up to 232 devices within a home or small business environment at a range up to 328 feet. Each smart home network has a unique network identification and each device within the home has node identification. The node identification is provided using the IPv6 address space. This provides for non-interference between smart configured homes within a neighborhood [z-wavealliance.org].

2.13   INDUSTRIAL INTERNET OF THINGS (IIoT)

IoT technology has found its way into a number of industries, as shown in Figure 2.16. This merger of IoT concepts and processes applied to industry has resulted in the IIoT. As an end of chapter assignment, we ask you to investigate one of these areas.

2.14   IoT AND IIoT SECURITY

IoT and IIoT security borrows many of the same measures from the IT world discussed earlier in the chapter. In addition, the International Society of Automation (ISA) and the International Electrotechnical Commission (IEC) have jointly developed a suite of security processes, procedures, and standards for control systems, as shown in Figure 2.17 [www.isa.org]. The literature contains documentation of nefarious actors penetrating a secure system. Typically, these attacks have breached the "air gap" and vulnerabilities between the IT and OT components of the system.

---

[3]AES 128 is a data encryption standard.

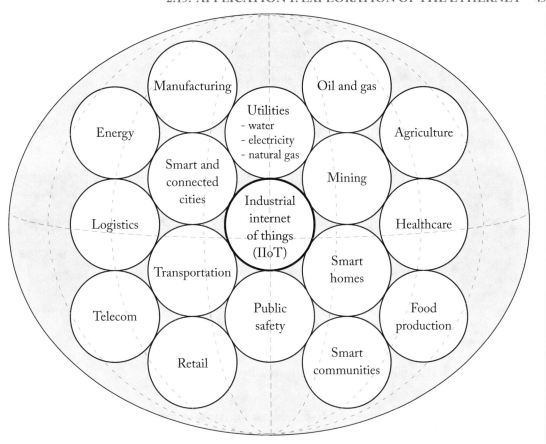

Figure 2.16: IIoT applications.

2.15 APPLICATION 1: EXPLORATION OF THE ETHERNET

In this exercise we use a Seeed Studio (www.seeedstudio.com) ethernet shield to communicate via the ethernet, as shown in Figure 2.18. The ethernet is a method of connecting computers within an LAN.

As shown in Figure 2.18, the ethernet shield is hosted by an Arduino UNO R3. The UNO R3 is in turn hosted by a laptop or PC equipped with the Arduino IDE. The ethernet shield is connected to a home router via an RJ45 cable. Another laptop/PC equipped with Telnet is also connected to the router via an RJ45 cable. The Telnet protocol provides for the connection to a remote computer via the internet.

To configure the ethernet shield, both the MAC address and the IP address assigned to the shield is needed. The MAC address of the shield is typically provided with the shield. To

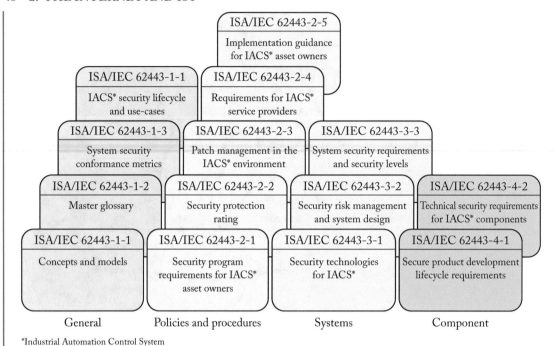

General   Policies and procedures   Systems   Component

*Industrial Automation Control System

Figure 2.17: ISA/IEC 62443 control system security [www.isa.org].

determine the IP address, use the DHCP Address Printer sketch available within the Arduino IDE example bank under Ethernet. The sketch needs to be updated with the shield's MAC address. When uploaded to the UNO R3, the IP address assigned is provided in the serial monitor.

To establish a two-way link over the ethernet, the Chat Server sketch is used. This sketch needs to be updated with both the MAC and IP address of the shield. The sketch is then uploaded to the UNO R3. The second PC/laptop should be equipped with Telnet. Telnet is a communication application readily available within Windows or through another application. For this exercise, PuTTY was used. It is an open-source terminal emulator application hosting Telnet and other communication applications. PuTTY is available for download from www.putty.org. Once PuTTY is started, the Telnet application is selected and the IP address of the ethernet shield is provided to the application. Two-way communication between the two devices then starts. Extensions:

1. **Extension 1.** Modify the sketch to send messages placed in the Arduino IDE serial monitor window to Telnet. Equip the UNO R3 with an LM34 Precision Fahrenheit Temperature Sensor. Send the sensed temperature to Telnet.

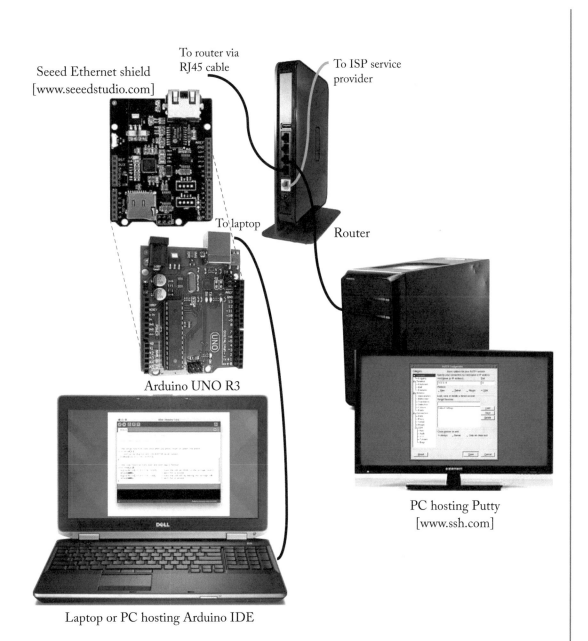

Figure 2.18: Ethernet configuration. Figures used by courtesy of Seeed Studio [www.seeedstudio.com] and the Arduino Team (CC BY-NC-SA) [www.arduino.cc].

2. **Extension 2.** Equip the second PC/laptop with an ethernet shield. Establish two-way communication between the two devices.

3. **Extension 3.** Equip the second PC/laptop with a WiFi shield. Establish two-way communication between the two devices.

2.16   APPLICATION 2: EXPLORATION WITH THE ARDUINO OPLÀ IoT KIT

Following the long-standing and admirable Arduino concept of open-source hardware and software, the Arduino Oplà Kit is an open platform to provide user-friendly access to the Arduino IoT Cloud. The kit features the Arduino MKR WiFi 1010 system on a chip (SOC) within a MKR IoT Carrier. The IoT Carrier is equipped with a OLED display; capacitive touch buttons; on-board sensors for temperature, humidity, pressure, and light; and an inertial measurement unit (IMU) as shown in Figure 2.19 [Arduino.cc].

MKR WiFi 1010 processor.   The MKR WiFi 1010 features a 32-bit ATSAMD21-M0 processor operating at 48 MHz. It is a full-featured processor with 256 KB of flash and 32 KB of RAM. The processor hosts SPI, I2C, and UART communication systems. The processor is also equipped with pulse width modulation (PWM) features and multiple channels of analog-to-digital conversion channels and a single digital-to-analog channel [opla.arduino.cc].

Exploring the Oplà IoT Kit.   To get acquainted with Oplà IoT Kit, go to opla.arduino.cc and complete these two exercises:

- "Get to Know the Carrier" and

- "Get to Know the Cloud."

Several notes on getting started:

- Ensure the "Arduino Web Editor Plug In" is properly installed. My virus checker tried to block installation. I had to temporarily suspend real time scanning by the virus checker to install the plug in.

- In the "Get to Know the Carrier" tutorial there is excellent sample sketch provided. Rather than type the entire program, the copy (Ctrl-C) and paste (Ctrl-V) features may be used to copy code snippets from the tutorial to assemble the entire program.

Once completed, consider completing one or more of the following step-by-step documented projects [opla.arduino.cc]:

- remote controlled lights,

- personal weather station,

(a) Arduino OPLA IoT Kit

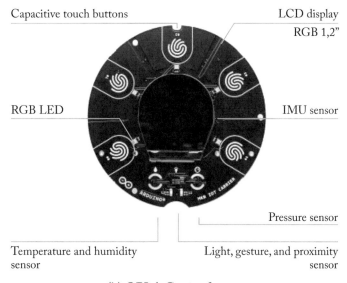

Capacitive touch buttons

LCD display
RGB 1,2"

RGB LED

IMU sensor

Pressure sensor

Temperature and humidity sensor

Light, gesture, and proximity sensor

(b) OPLA Carrier features

Figure 2.19: Arduino OPLÀ Kit. Arduino illustrations used with permission of the Arduino Team (CC BY-NC-SA) [arduino.cc].

- home security alarm,

- solar system tracker,

- inventory control,

- smart garden,

- thermostat control, or

- thinking about you.

2.17   APPLICATION 3: EXPLORATION OF THE MKR WiFi 1010 WITH THE ETHERNET

In this exercise we use the MKR WiFi 1010 from the Oplà Kit. The MKR WiFi 1010 may be left in the Oplà Carrier for this activity. The MKR 1010 is equipped with a WiFi module. We use the module to connect to your local ethernet using the WiFiNINA_Generic library. The library is available through the Arduino IDE Library Manager.

There are many example sketches within the library. I would recommend trying each of these:

- ScanNetworks

- WiFiPing

- WiFiChatServer

The sketches may require the name and password of your WiFi network.

To establish a two-link over the ethernet, the WiFiChatServer sketch is used. As in Application 1, a second PC/laptop equipped with Telnet is required. Telnet is a communication application readily available within Windows or through another application. For this exercise, PuTTY is used. It is an open-source terminal emulator application hosting Telnet and other communication applications. Once PuTTY is started, the Telnet application is selected and the IP address of the MKR 1010 is provided to the application. Two-way communication between the two devices then starts.

2.18   APPLICATION 4: EXPLORATION OF THE ARDUINO UNO WiFi REV 2 WITH THE ETHERNET

Arduino manufacturers an UNO R3 compatible microcontroller equipped with WiFi via a u-blox-NINA-W102 radio module, as shown in Figure 2.20. This board is called the Arduino UNO WiFi Rev 2. The host microcontroller on the board is the Microchip ATmega4809. In

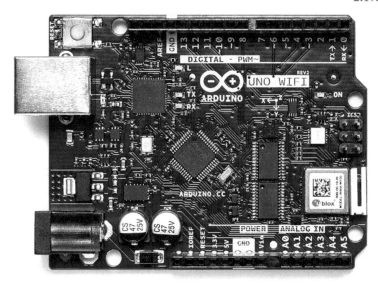

Figure 2.20: Arduino UNO WiFi Rev 2 Arduino illustrations used with permission of the Arduino Team (CC BY-NC-SA) [arduino.cc].

addition to the radio module it is equipped with an onboard IMU and a Bluetooth module (discussed in Chapter 3). In this application exercise, re-accomplish the activities of Application 3 above. **Note:** You will need to use the Arduino IDE Board Manager to install the "Arduino megaAVR Boards" package prior to accomplishing these activities [www.arduino.cc].

2.19 SUMMARY

The goal of this chapter was to provide a brief overview of internet concepts related to our exploration of the IoT. We began with a big picture overview of the internet followed by a brief review of internet development. We then took a closer look at selected internet topics including protocols, addressing techniques, different network configurations, threats to internet security, and threat countermeasures. We then investigated ethernet technology and the Arduino Oplà IoT Kit.

2.20 REFERENCES

[1] Arduino homepage. www.arduino.cc

[2] Bernstein, J. (2018). *Networking Made Easy—Get Yourself Connected*, James Bernstein.

[3] Colbach, G. (2019). *Wireless Networking—Introduction to Bluetooth and WiFi*, Gordon Colbach.

[4] Greer, C., Burns, M., Wollman, D., and Griffor, E. (2019). *Cyber—Physical Systems and Internet of Things*, NIST Special Publications 1900–202, National Institute of Standards and Technology, U.S. Department of Commerce. DOI: 10.6028/nist.sp.1900-202. 38, 39

[5] Hanes, D., Salgueiro, G., Grossetete, P., Barton, R., and Henry, J. (2017). *IoT Fundamentals—Networking Technologies, Protocols, and Use Cases for the Internet of Things*, Cisco Press. xvi, 24, 31, 33, 34, 38, 39, 40, 42, 43

[6] Internet Corporation for Assigned Names and Numbers (ICANN). www.ICANN.org

[7] Internet Engineering Task Force. www.IETF.org

[8] Kurose, J. and Ross, K. (1997). *Computer Networks—A Top-Down Approach*, 7th ed., Pearson Education, Inc. 35, 37

[9] Leiden, C. and Wilensky, M. (2009). *TCP/IP for Dummies*, 6th ed., John Wiley & Sons Publishing, Inc. 28, 29, 30

[10] Levine, R. and Levine Young, M. (2015). *The Internet for Dummies*, John Wiley & Sons Publishing, Inc. 26, 35, 37

[11] LOON homepage, *Connect People Everywhere*. www.loon.com 26

[12] Lowe, D. (2018). *Networking All-In-One for Dummies*, 7th ed., John Wiley & Sons Publishing, Inc. 28, 29, 31, 33, 34, 35, 37

[13] Microchip. www.microchip.com

[14] Null, L. and Lobur, J. (2015). *Computer Organization and Architecture*, Jones and Bartlett Learning. 28, 29, 30, 31, 33, 34

[15] Rajkumar, R., Lee, I., Sha, L., and Stankovic, J. (2010). *Cyber—Physical Systems: The Next Coupling Revolution*, ACM Design Automation Conference, Anaheim, CA. 38

[16] Shuler, R. (2002). *How does the Internet Work?*, Rus Shuler @ Pomeroy IT Solutions. 34, 35

[17] Stenerson, J. (2004). *Fundamentals of Programmable Logic Controllers, Sensors, and Communications*, Pearson Prentice Hall. 41, 42

[18] Z-Wave Alliance, *The Smart Home is Powered by Z-Wave*. z-wavealliance.org

2.21   CHAPTER PROBLEMS

1. What is meant by the internet cloud?

2. What are the differences between a network configuration for a home vs. a large business?

3. What is an ISP?

4. Draw the schematic of a home internet configuration using the Cisco System icons.

5. Draw the schematic of a large business internet configuration using the Cisco System icons.

6. In your own words, write a brief summary of historical network development.

7. What is the difference between ARPANET, DARPANET, and NSFNET?

8. Summarize the differences between the ISO/OSI and the TCP/IP layered protocol models.

9. Describe the difference in how IPv4 and IPv6 addresses are allocated.

10. What is CIDR addressing? How does it extend the IPv4 addressing space?

11. What is the mission of the following agencies: ICANN, IETF?

12. What is a MAC address? How is it different than an IP address?

13. What is the relationship between a DNS and an URL address?

14. Describe different sources of cybersecurity threats.

15. Describe measures to counter cybersecurity threats.

16. Provide a working definition of IoT and IIoT.

17. What is the difference between IT and OT? How are the concepts related?

18. What is a PLC?

19. Describe the PLC scanning process.

20. Provide an IoT model. Describe the interaction between things and applications.

21. What is an "air gap?" Why is it essential for IIoT security?

22. Research and write a short paper on an IIoT security breach. How was the system penetrated? How could have the situation been prevented?

CHAPTER 3

Connectivity

Objectives: After reading this chapter, the reader should be able to do the following.

- Describe the differences between serial and parallel communication.

- Provide definitions for key serial communications terminology.

- Describe how Near Field Communication (NFC) concepts are use for short range communications.

- Describe the operation of the Universal Synchronous and Asynchronous Serial Receiver and Transmitter (USART).

- Program the USART for basic transmission and reception using C.

- Describe the operation of the Serial Peripheral Interface (SPI).

- Program the SPI system using C.

- Describe the purpose of the Two Wire Interface (TWI).

- Program the TWI system using C.

- Describe the basic concepts supporting Bluetooth communications.

- Equip a microcontroller with a Bluetooth communication link.

- Describe the basic concepts supporting Zigbee communications.

- Equip a microcontroller with a Zigbee communication link.

- Equip a microcontroller with a cellular phone connection.

3.1   OVERVIEW

This chapter describes methods of connecting a microcontroller to external peripheral devices and other microcontrollers. Figure 3.1 provides a summary of network types and connection technologies to implement these networks. These concepts are important for our ongoing IoT exploration. For each range[1] we provide the network type and also implementation technologies.

---

[1]Note the range scale is not linear but a logarithm scale.

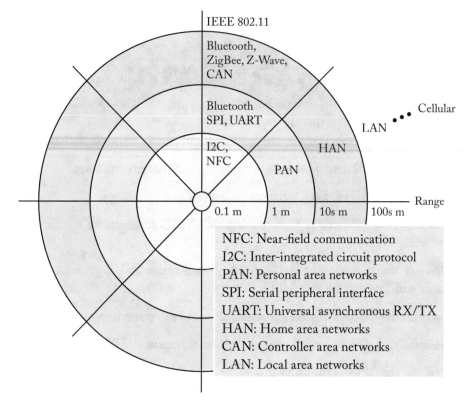

Figure 3.1: Connectivity (Hanes).

We explore some of these technologies in this chapter. The chapter begins with a brief description of serial communications and related serial communication terminology. We then review different connectivity technologies beginning with close range technologies, then mid-range technologies, then RF technologies, and conclude with GSM cellular phone technology.

3.2   SERIAL COMMUNICATIONS

Serial communication techniques provide a vital link between a microcontroller and certain input devices, output devices, and other microcontrollers. In this chapter, we investigate the serial communication features beginning with a review of serial communication concepts and terminology.[2] We then investigate in turn the following serial communication systems available on the ATmega328 microcontroller: the USART, the SPI, and the TWI. We provide guidance on how to program the USART, SPI, and TWI systems using the C programming language.[3]

---

[2]The sections on serial communication theory were adapted with permission from S. F. Barrett and D. J. Pack, *Microcontroller Fundamentals for Engineers and Scientists*, Morgan & Claypool Publishers, 2006.

[3]Intructions for programming the Arduino UNO R3 in C are provided in the Appendix A.

Microcontrollers must often exchange data with other microcontrollers or peripheral devices. Data may be exchanged by using parallel or serial techniques. With parallel techniques, an entire byte of data is typically sent simultaneously from the transmitting device to the receiver device. While this is efficient from a time point of view, it requires eight separate lines for the data transfer.

In serial transmission, a byte of data is sent a single bit at a time. Once 8 bits have been received at the receiver, the data byte is reconstructed. While this is inefficient from a time point of view, it only requires a line (or two) to transmit the data.

The ATmega328 (UNO R3) is equipped with a host of different serial communication subsystems including the serial USART, the serial peripheral interface or SPI, and the Two-wire Serial Interface (TWI). What all of these systems have in common is the serial transmission of data. As we shall see, these systems allow connection and control of additional peripheral devices. Before discussing the different serial communication features aboard these processors, we review serial communication terminology.

3.3   SERIAL COMMUNICATION TERMINOLOGY

In this section, we review common terminology associated with serial communication.

Asynchronous vs. Synchronous Serial Transmission: In serial communications, the transmitting and receiving device must be synchronized to one another and use a common data rate and protocol. Synchronization allows both the transmitter and receiver to be expecting data transmission/reception at the same time. There are two basic methods of maintaining "sync" between the transmitter and receiver: asynchronous and synchronous.

In an asynchronous serial communication system, such as the USART aboard the ATmega328, framing bits are used at the beginning and end of a data byte. These framing bits alert the receiver that an incoming data byte has arrived and also signals the completion of the data byte reception. The data rate for an asynchronous serial system is typically much slower than the synchronous system, but it only requires a single wire between the transmitter and receiver.

A synchronous serial communication system maintains "sync" between the transmitter and receiver by employing a common clock between the two devices. Data bits are sent and received on the edge of the clock. This allows data transfer rates higher than with asynchronous techniques but requires two lines, data and clock, to connect the receiver and transmitter.

Baud rate: Data transmission rates are typically specified as a Baud or bits per second rate. For example, 9600 Baud indicates the data is being transferred at 9600 bits per second.

Full Duplex: Often serial communication systems must both transmit and receive data. To do both transmission and reception, simultaneously, requires separate hardware for transmission and reception. A single duplex system has a single complement of hardware that must be switched from transmission to reception configuration. A full duplex serial communication system has separate hardware for transmission and reception.

Non-return to Zero (NRZ) Coding Format: There are many different coding standards used within serial communications. The important point is the transmitter and receiver must use a common coding standard so data may be interpreted correctly at the receiving end. The Microchip ATmega328 uses a non-return to zero (NRZ) coding standard. In NRZ coding a logic one is signaled by a logic high during the entire time slot allocated for a single bit, whereas a logic zero is signaled by a logic low during the entire time slot allocated for a single bit.

The RS-232 Communication Protocol: When serial transmission occurs over a long distance additional techniques may be used to insure data integrity. Over long distances logic levels degrade and may be corrupted by noise. At the receiving end, it is difficult to discern a logic high from a logic low. The RS-232 standard has been around for some time. With the RS-232 standard (EIA-232), a logic one is represented with a −12 VDC level while a logic zero is represented by a +12 VDC level. Chips are commonly available (e.g., MAX232) that convert the 5 and 0 V output levels from a transmitter to RS-232 compatible levels and convert back to 5 V and 0 V levels at the receiver. The RS-232 standard also specifies other features for this communication protocol.

Parity: To further enhance data integrity during transmission, parity techniques may be used. Parity is an additional bit (or bits) that may be transmitted with the data byte. The ATmega328 employs a single parity bit. With a single parity bit, a single bit error may be detected. Parity may be even or odd. In even parity, the parity bit is set to one or zero such that the number of ones in the data byte including the parity bit is even. In odd parity, the parity bit is set to 1 or 0 such that the number of ones in the data byte including the parity bit is odd. At the receiver, the number of bits within a data byte including the parity bit are counted to insure that parity has not changed, indicating an error, during transmission.

ASCII: The American Standard Code for Information Interchange or ASCII is a standardized, seven bit method of encoding alphanumeric data. It has been in use for many decades, so some of the characters and actions listed in the ASCII table are not in common use today. However, ASCII is still the most common method of encoding alphanumeric data. The ASCII code is provided in Figure 3.2. For example, the capital letter "G" is encoded in ASCII as 0x47. The "0x" symbol indicates the hexadecimal number representation. Unicode is the international counterpart of ASCII. It provides standardized 16-bit encoding format for the written languages of the world. ASCII is a subset of Unicode. The interested reader is referred to the Unicode home page website, www.unicode.org, for additional information on this standardized encoding format.

In the remainder of this chapter we examine different methods of connecting microcontrollers to a variety of devices. These concepts are essential in the IoT world.

		Most significant digit							
		0x0_	0x1_	0x2_	0x3_	0x4_	0x5_	0x6_	0x7_
	0x_0	NUL	DLE	SP	0	@	P	`	p
	0x_1	SOH	DC1	!	1	A	Q	a	q
	0x_2	STX	DC2	"	2	B	R	b	r
	0x_3	ETX	DC3	#	3	C	S	c	s
	0x_4	EOT	DC4	$	4	D	T	d	t
Least significant digit	0x_5	ENQ	NAK	%	5	E	U	e	u
	0x_6	ACK	SYN	&	6	F	V	f	v
	0x_7	BEL	ETB	'	7	G	W	g	w
	0x_8	BS	CAN	(	8	H	X	h	x
	0x_9	HT	EM	)	9	I	Y	i	y
	0x_A	LF	SUB	*	:	J	Z	j	z
	0x_B	VT	ESC	+	;	K	[	k	{
	0x_C	FF	FS	'	<	L	\	l		
	0x_D	CR	GS	-	=	M	]	m	}
	0x_E	SO	RS	.	>	N	^	n	~
	0x_F	SI	US	/	?	O	_	o	DEL

Figure 3.2: ASCII Code. The ASCII code is used to encode alphanumeric characters. The "0x" indicates hexadecimal notation in the C programming language.

3.4 NEAR FIELD COMMUNICATION

Near Field Communication (NFC) is closely related to Radio Frequency Identification (RFID). NFC allows the collection or exchange of data between two devices within 10 cm of one another. NFC uses a 13.56 MHz frequency to link the source (reader) with a second device. NFC may be employed in three different modes:

- a peer-to-peer link between two active (powered) devices,

- a card mode where the NFC device serves as a credit card substitute, or

- in a read/write mode.

In the read/write mode, an active (powered) reader is used to probe a passive (non-powered) tag. The tag contains the payload information. The link between the reader and tag is based on the concept of inductive coupling, as shown in Figure 3.3 (Maxim [12] and Triggs [19]).

As shown in Figure 3.3a, a coil of wire with a time varying current produces a magnetic field. If the coil is wrapped around a low reluctance path such as an iron core, the magnetic field

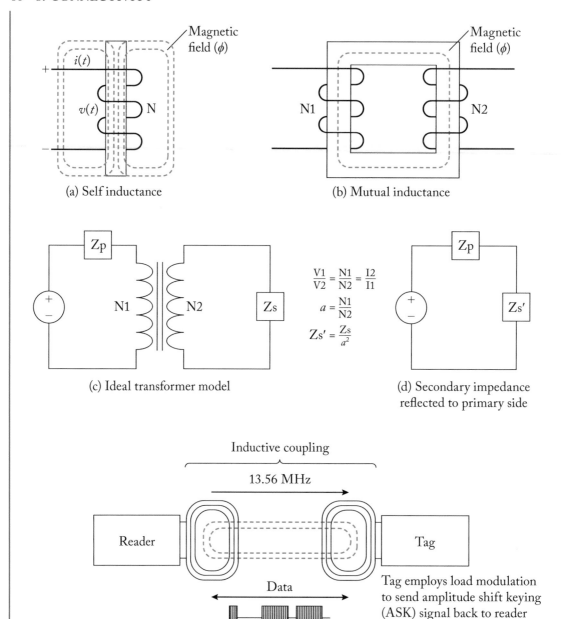

(a) Self inductance

(b) Mutual inductance

$$\frac{V1}{V2} = \frac{N1}{N2} = \frac{I2}{I1}$$

$$a = \frac{N1}{N2}$$

$$Zs' = \frac{Zs}{a^2}$$

(c) Ideal transformer model

(d) Secondary impedance reflected to primary side

Inductive coupling

13.56 MHz

Reader

Tag

Data

| 0 | 1 | 0 |

Tag employs load modulation to send amplitude shift keying (ASK) signal back to reader

(e) Simplified near field communication (NFC) model

Figure 3.3: **Near Field Communications** (Nilsson [15], Maxim [12], and Triggs [19]).

 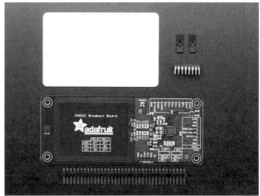

PN532 NFC/RFID Controller Shield for Arduino (#789) and PN532 NFC/RFID controller breakout board (#364)

Figure 3.4: Adafruit NFC products [www.adafruit.com].

is concentrated within the core and radiates around it. If the low reluctance path is configured as shown in Figure 3.3b, a transformer may be formed. The magnetic field from the primary coil with N1 turns is linked to the secondary coil with N2 turns via the magnetic field. The mutual inductance allows transformer actions such as voltage and current transformations, as shown in Figure 3.3c, based on the turns ratio between the primary and secondary. The model of an ideal transformer is shown in Figure 3.3c. As shown in Figure 3.3d, the secondary load impedance (Zs) has an impact on the transformer primary as a reflected impedance (Zs') (Nilsson [15]).

NFC employs these concepts to allow an active (powered) reader to probe a passive (non-powered) tag, as shown in Figure 3.3e. In NFC the reader coils are linked with the coils of the tag via a signal at a frequency of 13.56 MHz. The passive tag is powered by this signal. The tag delivers its payload of data back to the reader via load modulation. The series of ones and zeroes forming the data payload, turn a switch on or off consistent with the data. The switch controls the load impedance and hence the signal reflected back to the reader. Amplitude shift keying (ASK) modulation is used to encode the ones and zeros. The NFC data rate ranges from 106–424 kbps (Triggs [19], Maxim [12]).

Adafruit manufacturers several products including a shield and breakout board to explore NFC concepts. The boards may be used with a variety of Arduino-based products via a SPI or I2C connection. We revisit these boards in the application section of the chapter.

3.5   SERIAL USART

The serial USART[4] (or Universal Synchronous and Asynchronous Serial Receiver and Transmitter) provide for full-duplex (two-way) communication between a receiver and transmitter. This is accomplished by equipping the ATmega328 with independent hardware for the transmitter and receiver. The ATmega328 is equipped with a single USART channel. The USART is typically used for asynchronous communication. That is, there is not a common clock between the transmitter and receiver to keep them synchronized with one another. To maintain synchronization between the transmitter and receiver, framing start and stop bits are used at the beginning and end of each data byte in a transmission sequence. The Microchip USART also has synchronous features. Space does not permit a discussion of these USART enhancements.

The ATmega USART is quite flexible. It has the capability to be set to a variety of data transmission or Baud (bits per second) rates. The USART may also be set for data bit widths of 5–9 bits with one or two stop bits. Furthermore, the ATmega USART is equipped with a hardware generated parity bit (even or odd) and parity check hardware at the receiver. A single parity bit allows for the detection of a single bit error within a byte of data. The USART may also be configured to operate in a synchronous mode. We now discuss the operation, programming, and application of the USART. Due to space limitations, we cover only the most basic capability of this flexible and powerful serial communication system.

3.5.1   SYSTEM OVERVIEW

The block diagram for the USART is provided in Figure 3.5. The block diagram may appear a bit overwhelming but realize there are four basic pieces to the diagram: the clock generator, the transmission hardware, the receiver hardware, and three control registers (UCSRA, UCSBR, and UCSRC). We discuss each in turn.

USART Clock Generator

The USART Clock Generator provides the clock source for the USART system and sets the Baud rate for the USART. The Baud Rate is derived from the overall microcontroller clock source. The overall system clock is divided by the USART Baud rate Registers UBRR[H:L] and several additional dividers to set the Baud rate. For the asynchronous normal mode (U2X bit = 0), the Baud Rate is determined using the following expression:

$$Baud\ rate = (system\ clock\ frequency)/(16(UBRR + 1)),$$

where UBRR is the contents of the UBRRH and UBRRL registers (0 to 4095). Solving for UBRR yields:

$$UBRR = ((system\ clock\ generator)/(16 \times Baud\ rate)) - 1.$$

[4]The sections on USART, SPI, and TWI were adapted with permission from *Arduino II:Systems*, S.F. Barrett, Morgan & Claypool Publishers, 2020.

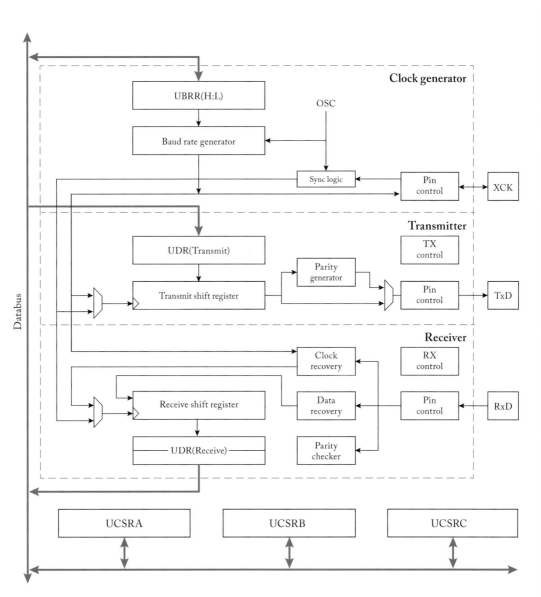

Figure 3.5: Microchip AVR ATmega USART block diagram. (Figure used with permission of Microchip, Incorporated [www.microchip.com].)

USART Transmitter

The USART transmitter consists of a Transmit Shift Register. The data to be transmitted is loaded into the Transmit Shift Register via the USART I/O Data Register (UDR). The start and stop framing bits are automatically appended to the data within the Transmit Shift Register. The parity is automatically calculated and appended to the Transmit Shift Register. Data is then shifted out of the Transmit Shift Register via the TxD pin a single bit at a time at the established Baud rate. The USART transmitter is equipped with two status flags: the UDRE and the TXC. The USART Data Register Empty (UDRE) flag sets when the transmit buffer is empty indicating it is ready to receive new data. This bit should be written to a zero when writing the USART Control and Status Register A (UCSRA). The UDRE bit is cleared by writing to the USART I/O Data Register (UDR). The Transmit Complete (TXC) Flag bit is set to logic one when the entire frame in the Transmit Shift Register has been shifted out and there are no new data currently present in the transmit buffer. The TXC bit may be reset by writing a logic one to it.

USART Receiver

The USART Receiver is virtually identical to the USART Transmitter except for the direction of the data flow is reversed. Data is received a single bit at a time via the RxD pin at the established Baud Rate. The USART Receiver is equipped with the Receive Complete (RXC) Flag. The RXC flag is logic one when unread data exists in the receive buffer.

USART Registers

In this section, we discuss the register settings for controlling the USART system. We have already discussed the function of the USART I/O Data Register (UDR) and the USART Baud Rate Registers (UBRRH and UBRRL). **Note:** The USART Control and Status Register C (UCSRC) and the USART Baud Rate Register High (UBRRH) are assigned to the same I/O location in the memory map. The URSEL bit (bit 7 of both registers) determine which register is being accessed. The URSEL bit must be 1 when writing to the UCSRC register and 0 when writing to the UBRRH register.

   Note: As previously mentioned, the ATmega328 is equipped with a single USART. The registers to configure the ATmega328 is provided in Figure 3.6.

USART Control and Status Register A (UCSRA)   The UCSRA register contains the RXC, TXC, and the UDRE bits. The function of these bits have already been discussed.

USART Control and Status Register B (UCSRB)   The UCSRB register contains the Receiver Enable (RXEN) bit and the Transmitter Enable (TXEN) bit. These bits are the "on/off" switch for the receiver and transmitter, respectively. The UCSRB register also contains the UCSZ2 bit. The UCSZ2 bit in the UCSRB register and the UCSZ[1:0] bits contained in the UCSRC register together set the data character size.

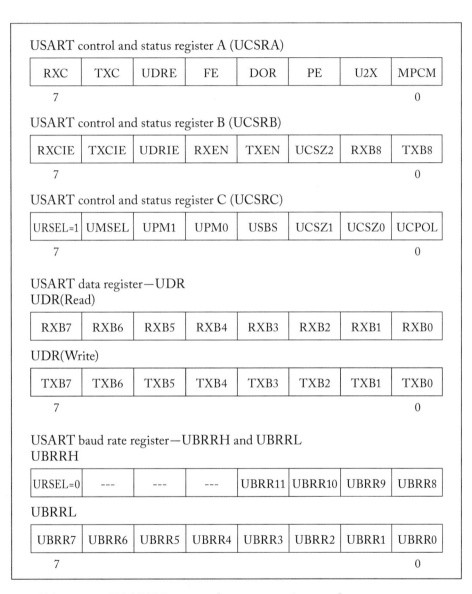

Figure 3.6: **ATmega328 USART Registers** [www.microchip.com].

USART Control and Status Register C (UCSRC)   The UCSRC register allows the user to customize the data features to the application at hand. It should be emphasized that both the transmitter and receiver be configured with the same data features for proper data transmission. The UCSRC contains the following bits:

- USART Mode Select (UMSEL) – 0: asynchronous operation, 1: synchronous operation,

- USART Parity Mode (UPM[1:0]) – 00: no parity, 10: even parity, 11: odd parity,

- USART Stop Bit Select (USBS) – 0: 1 stop bit, 1: 2 stop bits, and

- USART Character Size (data width) (UCSZ[2:0]) - 000: 5-bit, 001: 6-bit, 010: 7-bit, 011: 8-bit, 111: 9-bit.

3.5.2   PROGRAMMING IN ARDUINO

The Arduino Development Environment's Serial library provides over 20 functions to support USART operations. The functions include begin( ), end( ), print( ), println( ), read( ), and write( ).

In this example a Sparkfun LCD-09395, 5.0 VDC, serial, 16 by 2 character, black on white LCD display is connected to the Arduino UNO R3. Communication between the UNO R3 and the LCD is accomplished by a single 9600 bits per second (BAUD) connection.

Rather than use the onboard Universal Asynchronous Receiver Transmitter (USART), the Arduino Software Serial Library is used. The library provides functions to mimic USART activities on a digital pin. Details on the library are provided at the Arduino website [www. arduino.cc].

```
//*****************************************************************
//Example uses the Arduino Software Serial Library with the
//Sparkfun LCD-09395.
// - provides software-based serial port
//*****************************************************************

#include <SoftwareSerial.h>

//Specify Arduino pins for Serial connection:
//  SoftwareSerial LCD(RX_pin, TX_pin);
SoftwareSerial LCD(10, 11);

void setup()
{
```

```
LCD.begin(9600);                    //Baud rate: 9600 Baud
delay(500);                         //Delay for display
}

void loop()
{
//Cursor to line one, character one
LCD.write(254);                     //Command prefix
LCD.write(128);                     //Command

//clear display
LCD.write("                ");
LCD.write("                ");

//Cursor to line one, character one
LCD.write(254);                     //Command prefix
LCD.write(128);                     //Command

LCD.write("SerLCD Test");

//Cursor to line two, character one
LCD.write(254);                     //Command prefix
LCD.write(192);                     //Command

LCD.write("LCD-09395");

while(1);                           //pause here
}

//********************************************************************
```

3.5.3   SYSTEM OPERATION AND PROGRAMMING IN C

The basic activities of the USART system consist of initialization, transmission, and reception. These activities are summarized in Figure 3.7. Both the transmitter and receiver must be initialized with the same communication parameters for proper data transmission. The transmission and reception activities are similar except for the direction of data flow. In transmission, we monitor for the UDRE flag to set indicating the data register is empty. We then load the data for transmission into the UDR register. For reception, we monitor for the RXC bit to set in-

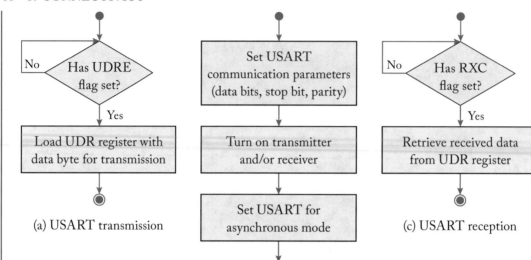

(a) USART transmission

(b) USART initialization

(c) USART reception

Figure 3.7: USART activities.

dicating there is unread data in the UDR register. We then retrieve the data from the UDR register.

Note: As previously mentioned, the ATmega328 is equipped with a single USART channel. The registers to configure the ATmega328 is provided in Figure 3.6.

To program the USART, we implement the flow diagrams provided in Figure 3.7. In the sample code provided, we assume the ATmega328 is operating at 10 MHz, and we desire a Baud Rate of 9600, asynchronous operation, no parity, one stop bit, and eight data bits.

To achieve 9600 Baud with an operating frequency of 10 MHz requires that we set the UBRR registers to 64 which is 0x40.[5]

```
//****************************************************************
//USART_init: initializes the USART system
//****************************************************************
```

---

[5]Chapter examples were originally developed for the ATmega164 and provided in *Microchip AVR Microcontroller Primer: Programming and Interfacing*, 3rd ed., S. Barrett and D. Pack, 2019. The examples were adapted with permission for the ATmega328.

```
void USART_init(void)
{
UCSRA = 0x00;                    //control register initialization
UCSRB = 0x08;                    //enable transmitter
UCSRC = 0x86;                    //async, no parity, 1 stop bit,
                                 // 8 data bits, Baud Rate initialization
UBRRH = 0x00;
UBRRL = 0x40;
}


//*********************************************************************
//USART_transmit: transmits single byte of data
//*********************************************************************

void USART_transmit(unsigned char data)
{
while((UCSRA & 0x20)==0x00)  //wait for UDRE flag
  {
  ;
  }
UDR = data;                      //load data to UDR for transmission
}


//*********************************************************************
//USART_receive: receives single byte of data
//*********************************************************************

unsigned char USART_receive(void)
{
while((UCSRA & 0x80)==0x00)  //wait for RXC flag
  {
  ;
  }
data = UDR;                      //retrieve data from UDR
return data;
}


//*********************************************************************
```

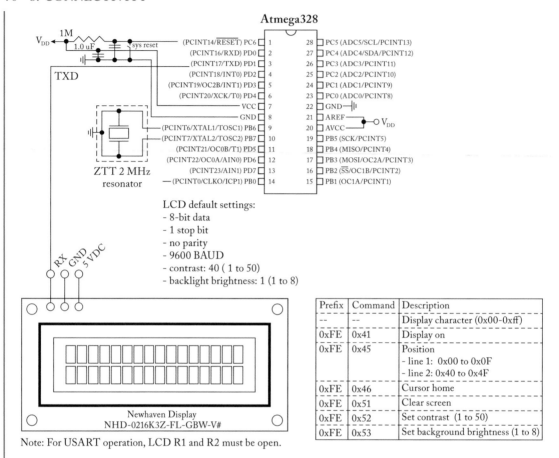

Note: For USART operation, LCD R1 and R2 must be open.

Figure 3.8: Serial LCD connections. ATmega328 diagram used courtesy of Microchip [www. microchip.com].

Example: Serial LCD

When developing embedded solutions, it is useful to receive status information from the micro-controller. Often liquid crystal displays (LCDs) are used for status display. LCDs are available in serial or parallel configuration. Serial LCDs communicate with the microcontroller via the USART system. In this example we configure the Newhaven Display #NHD-0216K3Z-FL-GBW-V3 to communicate with the ATmega328. The interface is shown in Figure 3.8. The ATmega328 USART channel is used. An abbreviated command set for the LCD is also shown. Characters are sent directly to the LCD; commands must be preceded by 0xFE.

In this specific example a 9600 BAUD rate is required by the LCD. The ATmega328 is clocked by a 2 MHz ceramic resonator. The UBRR registers (UBRR0H, UBRR0L) must be set to 12 (0x0c) to achieve the desired 9600 BAUD (bits per second rate). The ASCII representation of "G" is shown in Figure 3.9.

```
//**********************************************************************
//serial_LCD
//**********************************************************************

//Include Files: choose the appropriate include file depending on
//the compiler in use - comment out the include file not in use.

//include file(s) for JumpStart C for AVR Compiler****************
#include<iom328pv.h>                    //contains reg definitions

//include file(s) for the Atmel Studio gcc compiler
//#include <avr/io.h>                   //contains reg definitions

//function prototypes
void USART_init(void);
void USART_transmit(unsigned char data);
void LCD_init(void);
void lcd_print_string(char str[]);
void move_LCD_cursor(unsigned char position);

int main(void)
{
unsigned int i;

USART_init();
LCD_init();

while(1)
  {
  USART_transmit(0xFE);
  USART_transmit(0x46);                 //cursor to home

  for(i=0; i<10; i++)
    {
    USART_transmit('G');
    }
```

```
  //move cursor to line 2, position 0
  move_LCD_cursor(0x40);
  lcd_print_string("Test 1 - 2");
  }
}


//****************************************************************
//USART_init: initializes the USART system
//****************************************************************
void USART_init(void)
{
UCSR0A = 0x00;                       //control register init
UCSR0B = 0x08;                       //enable transmitter
UCSR0C = 0x06;                       //async, no parity,
                                     //1 stop bit, 8 data bits
                                     //Baud Rate initialization
UBRR0H = 0x00; UBRR0L = 0x0c;        //9600 BAUD, 2 MHz clock
                                     //divider set to 12

}


//****************************************************************
//USART_transmit: transmits single byte of data
//****************************************************************

void USART_transmit(unsigned char data)
{
while((UCSR0A & 0x20)==0x00) //wait for UDRE flag
  {
  ;
  }
UDR0 = data;                         //load data to UDR for tx
}


//****************************************************************
//LCD_init: initializes the USART system
//****************************************************************

void LCD_init(void)
```

```
{
USART_transmit(0xFE);
USART_transmit(0x41);                    //LCD on

USART_transmit(0xFE);
USART_transmit(0x46);                    //cursor to home
}


//*****************************************************************
//void lcd_print_string(char str[])
//*****************************************************************

void lcd_print_string(char str[])
{
int k = 0;

while(str[k] != 0x00)
  {
  USART_transmit(str[k]);
  k = k+1;
  }
}


//*****************************************************************
//void move_LCD_cursor(unsigned char position)
//*****************************************************************

void move_LCD_cursor(unsigned char position)
{
USART_transmit(0xFE);
USART_transmit(0x45);
USART_transmit(position);
}


//*****************************************************************
```

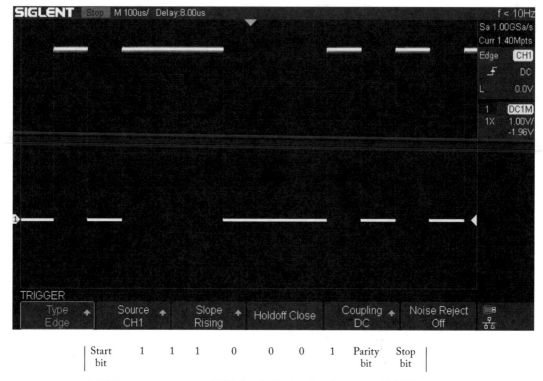

| Start bit | 1 | 1 | 1 | 0 | 0 | 0 | 1 | Parity bit | Stop bit |

ASCII representation of "G" (0x47, P100_0111) at 9600 BAUD.
The oscilloscope settings are 1 V/div horizontal, 100 us/div vertical.

Figure 3.9: ASCII representation of "G". Signal captured with a Siglent SDS 1104 X-E digital storage oscilloscope, four-channel, 100 MHz.

Example: PC Serial Monitor

During embedded system development, it is helpful to receive viewable status back from the microcontroller. Limited status may be sent to an LCD. In this example, we provide a one-way link between the ATmega328 and a support computer (PC or laptop). This allows considerable status to be sent and displayed on the support computer's monitor.

The signal from the microcontroller is 5 VDC (if a 5 VDC power supply is used). For proper interface to the PC, the 5 VDC signal must be translated to a compatible PC signal. This is easily accomplished using a USB cable with FTDI (Future Technology Devices International—www.ftdichip.com), set to 5 VDC. This cable is available from a number of sources. We use a Sparkfun Electronics (www.sparkfun.com) DEV-09718 illustrated in Figure 3.10. Driver installation instructions for the cable is provided at the Sparkfun website.

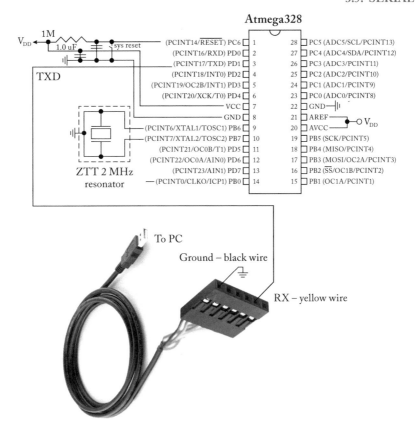

Figure 3.10: USART to computer communication link. Image used courtesy of Sparkfun (CC BY 2.0) [www.sparkfun.com]. ATmega328 diagram used courtesy of Microchip [www.microchip.com].

Messages sent from the ATmega328 are displayed on the support computer's monitor using a serial monitor program. In this example the open-source Arduino Software Integrated Development Environment (IDE) is used.

Provided below is a program illustrating how to send characters or messages from the ATmega328 to the support computer.

```
//********************************************************************
//usart_to_pc: provides one way communication from ATmega328
//USART0 back to a host (PC or laptop).
//In this example the ATmega328 was clocked by a 2.0 MHz
//external ceramic resonator.
```

```
//****************************************************************

//Include Files: choose the appropriate include file depending on
//the compiler in use - comment out the include file not in use.

//include file(s) for JumpStart C for AVR Compiler****************
#include<iom328pv.h>                    //contains reg definitions

//include file(s) for the Atmel Studio gcc compiler
//#include <avr/io.h>                   //contains reg definitions

//function prototypes
void USART_init(void);
void USART_transmit(unsigned char data);
void lcd_print_string(char str[]);
void delay_100ms(void);
void delay_1s(void);


int main(void)
{
unsigned int i;

USART_init();

for(i=0; i<5; i++)
  {
  USART_transmit('G');
  lcd_print_string("\n");
  delay_100ms();
  }

for(i=0; i<5; i++)
  {
  lcd_print_string("Test print\n");
  delay_100ms();
  }
```

```c
lcd_print_string("Test print\n\n");  //newline
lcd_print_string("Test \tprint\n");  //horizontal tab
}


//**********************************************************************
//delay_100ms: inaccurate, yet simple method of creating delay
//  - processor clock: ceramic resonator at 2.0 MHz
//  - 100 ms delay requires 200,000 clock cycles
//  - nop requires 1 clock cycle to execute
//**********************************************************************

void delay_100ms(void)
{
unsigned int i,j;

for(i=0; i < 200; i++)
  {
  for(j=0; j < 1000; j++)
    {
    asm("nop");
    }
  }
}


//**********************************************************************
//delay_1s: inaccurate, yet simple method of creating delay
//  - processor clock: ceramic resonator at 2.0 MHz
//  - 100 ms delay requires 200,000 clock cycles
//  - nop requires 1 clock cycle to execute
//  - call 10 times for 1s delay
//**********************************************************************

void delay_1s(void)
{
unsigned int i;

for(i=0; i< 10; i++)
  {
```

```
    delay_100ms();
    }
}


//****************************************************************
//USART_init: initializes the USART system
//****************************************************************

void USART_init(void)
{
UCSR0A = 0x00;                       //control register init
UCSR0B = 0x08;                       //enable transmitter
UCSR0C = 0x06;                       //async, no parity,
                                     //1 stop bit, 8 data bits
                                     //Baud Rate initialization
UBRR0H = 0x00; UBRR0L = 0x0c;        //9600 BAUD, 2 MHz clock
                                     //divider set to 12 (0x0c)

}


//****************************************************************
//USART_transmit: transmits single byte of data
//****************************************************************

void USART_transmit(unsigned char data)
{
while((UCSR0A & 0x20)==0x00) //wait for UDRE flag
    {
    ;
    }
UDR0 = data;                         //load data to UDR for tx
}


//****************************************************************
//LCD_init: initializes the USART system
//****************************************************************

void LCD_init(void)
{
```

```
USART_transmit(0xFE);
USART_transmit(0x41);                    //LCD on

USART_transmit(0xFE);
USART_transmit(0x46);                    //cursor to home
}


//******************************************************************
//void lcd_print_string(char str[])
//******************************************************************

void lcd_print_string(char str[])
{
int k = 0;

while(str[k] != 0x00)
  {
  USART_transmit(str[k]);
  k = k+1;
  }
}


//******************************************************************
//void move_LCD_cursor(unsigned char position)
//******************************************************************

void move_LCD_cursor(unsigned char position)
{
USART_transmit(0xFE);
USART_transmit(0x45);
USART_transmit(position);
}


//******************************************************************
```

3.6   SERIAL PERIPHERAL INTERFACE (SPI)

The ATmega Serial Peripheral Interface or SPI provides for two-way serial communication between a transmitter and a receiver. In the SPI system, the transmitter and receiver share a common clock source. This requires an additional clock line between the transmitter and receiver but allows for higher data transmission rates as compared to the USART. The SPI system allows for fast and efficient data exchange between microcontrollers or peripheral devices. There are many SPI compatible external systems available to extend the features of the microcontroller. For example, a liquid crystal display or a digital-to-analog converter could be added to the microcontroller using the SPI system.

3.6.1   SPI OPERATION

The SPI may be viewed as a synchronous 16-bit shift register with an 8-bit half residing in the transmitter and the other 8-bit half residing in the receiver, as shown in Figure 3.11. The transmitter is designated the master since it is providing the synchronizing clock source between the transmitter and the receiver. The receiver is designated as the slave. A slave is chosen for reception by taking its Slave Select ($\overline{SS}$) line low. When the $\overline{SS}$ line is taken low, the slave's shifting capability is enabled.

SPI transmission is initiated by loading a data byte into the master configured SPI Data Register (SPDR). At that time, the SPI clock generator provides clock pulses to the master and also to the slave via the SCK pin. A single bit is shifted out of the master designated shift register on the Master Out Slave In (MOSI) microcontroller pin on every SCK pulse. The data is received at the MOSI pin of the slave designated device. At the same time, a single bit is shifted out of the Master In Slave Out (MISO) pin of the slave device and into the MISO pin of the master device. After eight master SCK clock pulses, a byte of data has been exchanged between the master and slave designated SPI devices. Completion of data transmission in the master and data reception in the slave is signaled by the SPI Interrupt Flag (SPIF) in both devices. The SPIF flag is located in the SPI Status Register (SPSR) of each device. At that time, another data byte may be transmitted.

3.6.2   REGISTERS

The registers for the SPI system are provided in Figure 3.12. We will discuss each one in turn.

SPI Control Register (SPCR)
The SPI Control Register (SPCR) contains the "on/off" switch for the SPI system. It also provides the flexibility for the SPI to be connected to a wide variety of devices with different data formats. It is important that both the SPI master and slave devices be configured for compatible data formats for proper data transmission. The SPCR contains the following bits.

- SPI Enable (SPE) is the "on/off" switch for the SPI system. A logic one turns the system on and logic zero turns it off.

- Data Order (DORD) allows the direction of shift from master to slave to be controlled. When the DORD bit is set to one, the least significant bit (LSB) of the SPI Data

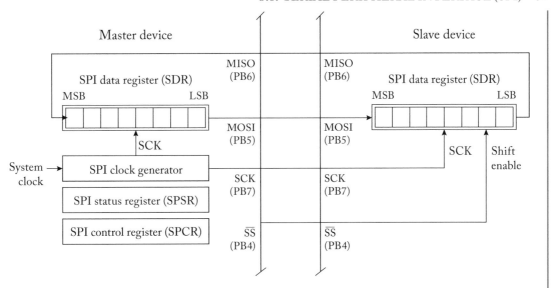

Figure 3.11: SPI overview.

SPI status register—SPSR

SPIE	SPE	DORD	MSTR	CPOL	CPHA	SPR1	SPR0
7							0

SPI control register—SPCR

SPIF	WCOL	---	---	---	---	---	SPI2X
7							0

SPI data register—SPDR

MSB							LSB
7							0

Figure 3.12: SPI registers [www.microchip.com].

Register (SPDR) is transmitted first. When the DORD bit is set to zero the Most Significant Bit (MSB) of the SPDR is transmitted first.

- The Master/Slave Select (MSTR) bit determines if the SPI system will serve as a master (logic one) or slave (logic zero).

- The Clock Polarity (CPOL) bit allows determines the idle condition of the SCK pin. When CPOL is one, SCK will idle logic high, whereas when CPOL is 0, SCK will idle logic 0.

- The Clock Phase (CPHA) determines if the data bit will be sampled on the leading (0) or trailing (1) edge of the SCK.

- The SPI SCK is derived from the microcontroller's system clock source. The system clock is divided down to form the SPI SCK. The SPI Clock Rate Select bits SPR[1:0] and the Double SPI Speed Bit (SPI2X) are used to set the division factor. The following divisions may be selected using SPI2X, SPR1, SPR0:

  - 000: SCK = system clock/4
  - 001: SCK = system clock/16
  - 010: SCK = system clock/64
  - 011: SCK = system clock/1284
  - 100: SCK = system clock/2
  - 101: SCK = system clock/8
  - 110: SCK = system clock/32
  - 111: SCK = system clock/64

SPI Status Register (SPSR)

The SPSR contains the SPIF. The flag sets when eight data bits have been transferred from the master to the slave. The SPIF bit is cleared by first reading the SPSR after the SPIF flag has been set and then reading the SPI Data Register (SPDR). The SPSR also contains the SPI2X bit used to set the SCK frequency.

SPI Data Register (SPDR)   As previously mentioned, writing a data byte to the SPDR initiates SPI transmission.

3.6.3   SPI PROGRAMMING IN THE ARDUINO DEVELOPMENT ENVIRONMENT

The ADE provides the "shiftOut" command to provide ISP style serial communications [www.arduino.cc]. The shiftOut command requires four parameters when called:

- **dataPin:** the Arduino UNO R3 DIGITAL pin to be used for serial output.

- **clockPin:** the Arduino UNO R3 DIGITAL pin to be used for the clock.

- **bitOrder:** indicates whether the data byte will be sent most significant bit first (MSBFIRST) or least significant bit first (LSBFIRST).

- **value:** the data byte that will be shifted out.

To use the shiftOut command, the appropriate pins are declared as output using the pinMode command in the setup() function. The shiftOut command is then called at the appropriate place within the loop() function using the following syntax:

```
shiftOut(dataPin, clockPin, LSBFIRST, value);
```

As a result of the this command, the value specified will be serially shifted out of the data pin specified, least significant bit first, at the clock rate provided at the clock pin.

3.6.4   SPI PROGRAMMING IN C

To program the SPI system in C, the system must first be initialized with the desired data format. Data transmission may then commence. Functions for initialization, transmission, and reception are provided below. In this specific example, we divide the clock oscillator frequency by 128 to set the SCK clock frequency. **Note:** For proper SPI operation the slave select pin (PB2) must be set to output even if not used [www.avrfreaks.com].

```
//****************************************************************
//spi_init: initializes spi system
//****************************************************************

void spi_init(unsigned char control)
{
DDRB = 0x2c;              //Set SCK (PB5), MOSI (PB3), /SS (PB2)
                         //for output, others to input
SPCR = 0x53;             //Configure SPI Control Register (SPCR)
                         //SPIE:0,SPE:1,DORD:0,MSTR:1
                         //CPOL:0,CPHA:0,SPR:1,SPR0:1
                         //Divide clock by 128
}


//****************************************************************
//spi_write: Used by SPI master to transmit a data byte
//****************************************************************

void spi_write(unsigned char byte)
{
SPDR = byte;
```

```
while (!(SPSR & 0x80));
}


//****************************************************************
//spi_read: Used by SPI slave to receive data byte
//****************************************************************


unsigned char spi_read(void)
{
while (!(SPSR & 0x80));


return SPDR;
}


//****************************************************************
```

3.6.5   EXAMPLE: LED STRIP

LED strips may be used for motivational (fun) optical displays, games, or for instrumentation-based applications. In this example we control an LPD8806-based LED strip. We use a one meter, 32 RGB LED strip available from Adafruit (#306) for approximately $30 USD [www.adafruit.com].

The red, blue, and green component of each RGB LED is independently set using an 8-bit code. The most significant bit (MSB) is logic one followed by 7 bits to set the LED intensity (0 to 127). The component values are sequentially shifted out of the ATmega328 using the SPI features. The first component value shifted out corresponds to the LED nearest the microcontroller. Each shifted component value is latched to the corresponding R, G, and B component of the LED. As a new component value is received, the previous value is latched and held constant. An extra byte is required to latch the final parameter value. A zero byte $(00)_{16}$ is used to complete the data sequence and reset back to the first LED [www.adafruit.com].

Only four connections are required between the ATmega328 and the LED strip as shown in Figure 3.13. The connections are color coded: red-power, black-ground, yellow-data, and green-clock. The ATmega328 is equipped with a single SPI channel. This channel is used to program the ATmega328 using ISP techniques. It is important to note the LED strip requires a supply of 5 VDC and a current rating of 2 amps per meter of LED strip. In this example we use the Adafruit #276 5V 2A (2000mA) switching power supply [www.adafruit.com].

In this example each RGB component is sent separately to the strip. The example illustrates how each variable in the program controls a specific aspect of the LED strip. Here are some important implementation notes.

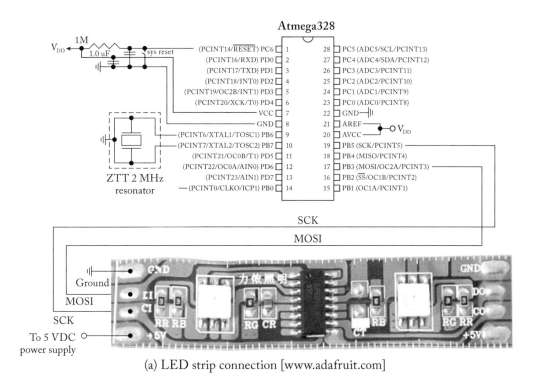

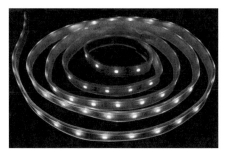

(a) LED strip connection [www.adafruit.com]

(b) LED strip by the meter [www.adafruit.com]

Figure 3.13: ATmega328-controlling LED strip [www.adafruit.com]. ATmega328 diagram used courtesy of Microchip [www.microchip.com].

- SPI must be configured for most significant bit (MSB) first.

- LED brightness is 7 bits. Most significant bit (MSB) must be set to logic one.

- Each LED requires a separate R-G-B intensity component. The order of data is G-R-B.

- After sending data for all LEDs. A byte of (0x00) must be sent to return strip to first LED.

- Data stream for each LED is: 1-G6-G5-G4-G3-G2-G1-G0-1-R6-R5-R4-R3-R2-R1-R0-1-B6-B5-B4-B3-B2-B1-B0.

```
//*****************************************************************
//spi.c
//*****************************************************************
//RGB led strip tutorial: illustrates different variables within
//RGB LED strip
//
//LED strip LDP8806 - available from www.adafruit.com (#306)
//
//Connections:
// - External 5 VDC supply - Adafruit 5 VDC, 2A (#276) - red
// - Ground - black
// - Serial Data In:  ATmega328: MOSI PORTB[3], pin 17
// -            CLK:  ATmega328: SCK  PORTB[5], pin 19
//
//Variables:
// - LED_brightness - set intensity from 0 to 127
// - segment_delay  - delay between LED RGB segments
// - strip_delay    - delay between LED strip update
//
//Notes:
// - SPI must be configured for Most significant bit (MSB) first
// - LED brightness is seven bits.  Most significant bit (MSB)
//    must be set to logic one
// - Each LED requires a separate R-G-B intensity components.
//    The order of data is G-R-B.
// - After sending data for all strip LEDs.  A byte of (0x00) must
//    be sent to return strip to first LED.
// - Data stream for each LED is:
```

```
//1-G6-G5-G4-G3-G2-G1-G0-1-R6-R5-R4-R3-R2-R1-R0-1-B6-B5-B4-B3-B2-B1-B0
//
//This example code is in the public domain.
//*****************************************************************

#define LED_strip_latch 0x00

//function prototypes
void spi_init(void);
void spi_write(unsigned char byte);
void delay_1s(void);
void delay_100ms(void);
void clear_strip(void);

//Include Files: choose the appropriate include file depending on
//the compiler in use - comment out the include file not in use.

//include file(s) for JumpStart C for AVR Compiler****************
#include<iom328pv.h>              //contains reg definitions

//include file(s) for the Atmel Studio gcc compiler
//#include <avr/io.h>             //contains reg definitions

unsigned char    strip_length = 32; //number of RGB LEDs in strip
unsigned char    LED_brightness;    //0 to 127
unsigned char    position;          //LED position in strip

int main(void)
{
spi_init();
spi_write(LED_strip_latch);        //reset to first segment
clear_strip();                     //all strip LEDs to black
delay_100ms();

//increment the green intensity of the strip LEDs
for(LED_brightness = 0; LED_brightness <= 60;
    LED_brightness = LED_brightness + 10)
    {
```

```
    for(position = 0; position<strip_length; position = position+1)
      {
      spi_write(0x80 | LED_brightness);   //Green - MSB 1
      spi_write(0x80 | 0x00);             //Red   - none
      spi_write(0x80 | 0x00);             //Blue  - none

      delay_100ms();
      }
    spi_write(LED_strip_latch);           //reset to first segment
    delay_100ms();
    }

clear_strip();                            //all strip LEDs to black
delay_100ms();

//increment the red intensity of the strip LEDs
for(LED_brightness = 0; LED_brightness <= 60;
    LED_brightness = LED_brightness + 10)
  {
  for(position = 0; position<strip_length; position = position+1)
    {
    spi_write(0x80 | 0x00);             //Green - none
    spi_write(0x80 | LED_brightness);   //Red   - MSB1
    spi_write(0x80 | 0x00);             //Blue  - none

    delay_100ms();
    }
  spi_write(LED_strip_latch);           //reset to first segment
  delay_100ms();
  }

clear_strip();                            //all strip LEDs to black
delay_100ms();

//increment the blue intensity of the strip LEDs
for(LED_brightness = 0; LED_brightness <= 60;
    LED_brightness = LED_brightness + 10)
    {
```

```
      for(position = 0; position<strip_length; position = position+1)
        {
        spi_write(0x80 | 0x00);          //Green - none
        spi_write(0x80 | 0x00);          //Red   - none
        spi_write(0x80 | LED_brightness); //Blue  - MSB1

        delay_100ms();
        }
      spi_write(LED_strip_latch);        //reset to first segment
      delay_100ms();
      }

clear_strip();                           //all strip LEDs to black
delay_100ms();

}


//****************************************************************

void clear_strip(void)
{
//clear strip
for(position = 0; position<strip_length; position = position+1)
  {
  spi_write(0x80 | 0x00);            //Green - none
  spi_write(0x80 | 0x00);            //Red - none
  spi_write(0x80 | 0x00);            //Blue  - none

  spi_write(LED_strip_latch);        //Latch with zero
  delay_100ms();                     //clear delay
  }
}

//****************************************************************
//spi_init: initializes spi system
//****************************************************************
void spi_init()
{
```

```
DDRB = 0x2c;                //Set SCK (PB5), MOSI (PB3), /SS (PB2)
                            //for output, others to input
                            //Configure SPI Control Register (SPCR)
SPCR = 0x5F;                //SPIE:0
                            //SPE: 1  SPI on
                            //DORD:0  MSB first
                            //MSTR:1  Master (provides clock)
                            //CPOL:1  Required by LED strip
                            //CPHA:1  Required by LED strip
                            //SPR1:1  SPR[1:0] 11: div clock 128
                            //SPR0:1
}


//*****************************************************************
//spi_write: Used by SPI master to transmit a data byte
//*****************************************************************

void spi_write(unsigned char byte)
{
SPDR = byte;
while (!(SPSR & 0x80))
   {
   ;
   }
}


//*****************************************************************
//delay_100ms: inaccurate, yet simple method of creating delay
// - processor clock: ceramic resonator at 2.0 MHz
// - 100 ms delay requires 200,000 clock cycles
// - nop requires 1 clock cycle to execute
//*****************************************************************

void delay_100ms(void)
{
unsigned int i,j;

for(i=0; i < 200; i++)
```

```
  {
  for(j=0; j < 1000; j++)
    {
    asm("nop");
    }
  }
}


//*****************************************************************
//delay_1s: inaccurate, yet simple method of creating delay
//  - processor clock: ceramic resonator at 2.0 MHz
//  - 100 ms delay requires 200,000 clock cycles
//  - nop requires 1 clock cycle to execute
//  - call 10 times for 1s delay
//*****************************************************************

void delay_1s(void)
{
unsigned int i;

for(i=0; i< 10; i++)
  {
  delay_100ms();
  }
}


//*****************************************************************
```

3.7   TWO-WIRE SERIAL INTERFACE

The TWI subsystem allows the system designer to connect a number of TWI configured devices (microcontrollers, transducers, displays, memory storage, etc.) together into a system using a two-wire interconnecting scheme. The TWI allows a maximum of 128 devices to be connected together. Each device has its own unique address and may both transmit and receive over the two-wire bus at frequencies up to 400 kHz. This allows the device to freely exchange information with other devices in a small area network. The TWI is alternately known as the Inter-Integrated Circuit (I^2C) protocol (Philips [11]).

An overview of the TWI system is shown in Figure 3.14. Devices within the small area network are connected by two wires to share data (SDA) and a common clock (SCL). Pullup

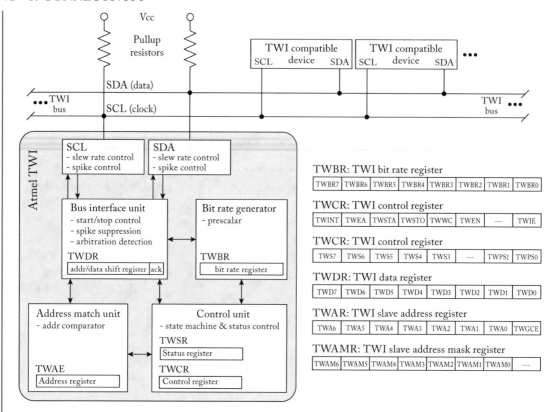

Figure 3.14: **TWI system overview** [www.microchip.com].

resistors are required on each of the lines. TWI compatible devices are connected to the SCL and SDA lines as shown.

The Microchip TWI system is a state machine to control the "hand shaking" protocol between the TWI master(s) and the multiple slave devices on the TWI bus. If the system contains more than one master designated device, arbitration detection and resolution protocols prevent bus contention. Each slave device has a unique seven bit address to allow one-to-one communication using the Address Match Unit and address comparator. The TWI bus frequency should not exceed 400 kHz. The bus frequency is derived from the Microchip microcontroller clock signal using the Bit Rate Generator which contains prescalar hardware. The Microchip TWI system also includes signal conditioning features for the SCL and SDA pins including slew rate and spike control.

The TWI system is configured and controlled using a series of registers shown in Figure 3.14. Details on specific bit settings are provided in the Microchip ATmega328 datasheet and will not be duplicated here [www.microchip.com]. These include:

- TWBR: TWI Bit Rate Register

- TWCR: TWI Control Register

- TWSR: TWI Status Register

- TWDR: TWI Data Register

- TWAR: TWI Address Register

- TWAMR: TWI Slave Address Mask Register

Data is exchanged by devices on the TWI bus using s carefully orchestrated "hand shaking" protocol as shown in Figure 3.15. On the left-hand side of the figure are the actions required by the TWI application program and the right side of the figure contains the response from the TWI compatible slave hardware. At each step in the exchange protocol action is initiated by the application program hosted on the TWI master device with a corresponding response from the slave configured device. If the expected response is not received, an error is triggered.

3.7.1   TWI PROGRAMMING ARDUINO DEVELOPMENT ENVIRONMENT

The Arduino Development Environment's Wire library provides ten functions to support TWI/I2C operations. The functions include begin(), beginTransmission(), endTransmission(), write(), read(), and SetClock(). We use these functions in the chapter's Application section.

3.7.2   TWI PROGRAMMING IN C–TWI COMPATIBLE LCD

In this example a TWI compatible LCD is used to display temperature data from an LM34 temperature sensor as shown in Figure 3.16. Note how the LCD is connected to the TWI bus via the SDA and SCL pins.

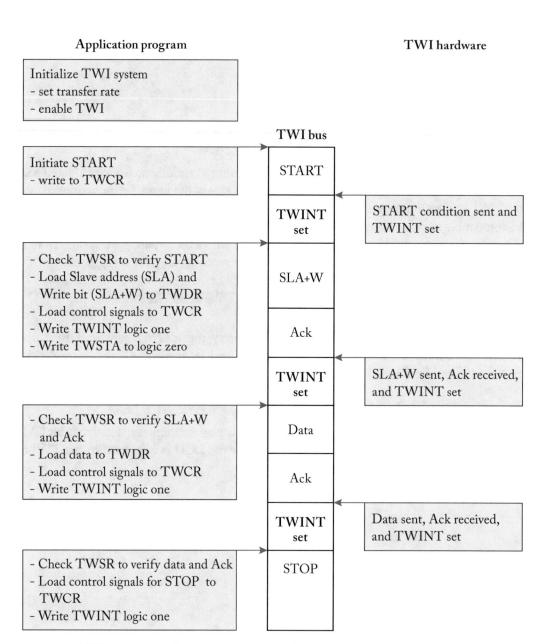

Figure 3.15: **TWI** operations [www.microchip.com].

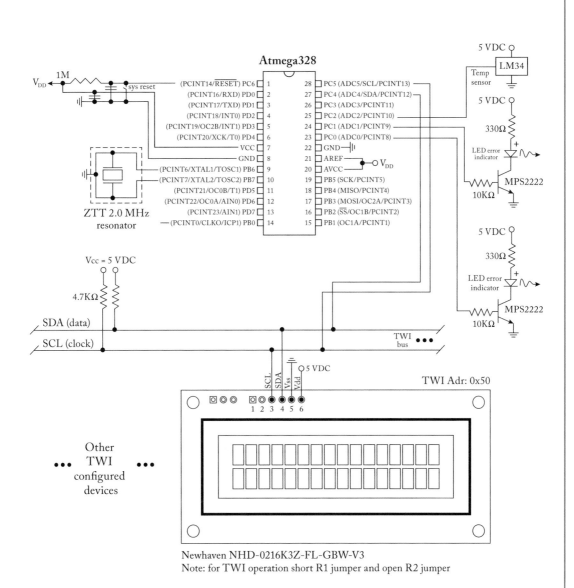

Figure 3.16: **TWI connecting sensor to LCD** [www.sparkfun.com]. ATmega328 diagram used courtesy of Microchip [www.microchip.com].

```
//*****************************************************************
//twi2.c
//
//ATmega328 is clocked by a 2.0 MHz ceramic resonator
//
//Example provides twi communication with twi configured LCD.
// - LCD: Newhaven NHD-0216K3Z-FL-GBW-V3
// - LCD short R1 jumper, open R2 jumper
//Adapted from Microchip provided TWI examples [www.microchip.com]
//*****************************************************************
//Include Files: choose the appropriate include file depending on
//the compiler in use - comment out the include file not in use.
//
//include file(s) for JumpStart C for AVR Compiler***************
#include<iom328pv.h>              //contains reg definitions

//include file(s) for the Atmel Studio gcc compiler
//#include <avr/io.h>             //contains reg definitions

//TWSR status codes with prescaler = 0
#define START           0x08    //START condition transmitted
#define START_REP       0x10    //Repeated START transmitted
#define MT_SLA_NO_ACK   0x20    //SLA+W has been transmitted,
                                //NOT ACK has been received
#define MT_SLA_ACK      0x18    //SLA+W has been transmitted,
                                //ACK has been received
#define MT_DATA_ACK     0x28    //Data byte has been transmitted,
                                //ACK has been received
#define MT_DATA_NO_ACK  0x30    //Data byte has been transmitted,
                                //NOT ACK has been received
#define MR_SLA_ACK      0x40    //SLA+R has been transmitted,
                                //ACK has been received
#define ARB_LOST        0x38    //Arbitration lost in SLA+W or
                                //data bytes

//clock specifications
#define ceramic_res_freq 2000000UL   //ATmega328 operating freq
#define scl_clock        100000L     //desired TWI bus freq
```

```
//peripheral device addresses
#define LCD_twi_addr       0x50     //addr - LSB 0 for write

//function prototypes
void initialize_ports(void);
void ERROR(unsigned char error_number);
void LCD_init(void);
void lcd_print_string(char str[]);
void move_LCD_cursor(unsigned char position);
void twi_initialize(void);
void twi_send_byte(unsigned char slave_device_addr,
                          unsigned char send_data);
void InitADC( void);
unsigned int ReadADC(unsigned char channel);
void temperatureToLCD(unsigned int);
void delay_10ms(void);
void delay_100ms(void);

void main(void)
{
unsigned int temp_int;

initialize_ports();
twi_initialize();
InitADC();
LCD_init();
delay_100ms();

while(1)
  {
  twi_send_byte(LCD_twi_addr, 0xFE);
  twi_send_byte(LCD_twi_addr, 0x51); //clear LCD
  delay_10ms();

  twi_send_byte(LCD_twi_addr, 0xFE);
  twi_send_byte(LCD_twi_addr, 0x46); //cursor to home
  lcd_print_string("Temp:");
  delay_10ms();
```

```
   temp_int = ReadADC(0x02);              //read temp from LM34

                                          //move cursor line 2, pos 0
   move_LCD_cursor(0x40);
   temperatureToLCD(temp_int);
   delay_100ms();
   delay_100ms();
   }
}


//*******************************************************************
//initialize_ports: provides initial configuration for I/O ports
//*******************************************************************

void initialize_ports(void)
{
DDRB=0xff;                         //PORTB[7:0] as output
PORTB=0x00;                        //initialize low

DDRC=0xfb;                         //set PORTC as output, C[2] input
PORTC=0x00;                        //initialize low

DDRD=0xff;                         //set PORTD as output
PORTD=0x00;                        //initialize low

}


//*******************************************************************
//void ERROR - indicates error source with LED pattern
//*******************************************************************

void ERROR(unsigned char error_number)
{
//Turn off error LEDs
PORTC = 0x00;

if(error_number == 1)              //Error 01
  PORTC = 0x01;
```

```
else if(error_number == 2)        //Error 10
  PORTC = 0x02;
else if(error_number == 3)        //Error 11
  PORTC = 0x03;
else
  PORTC = 0x00;
}


//****************************************************************
//LCD_init: initializes the USART system
//****************************************************************
void LCD_init(void)
{
twi_send_byte(LCD_twi_addr, 0xFE);
twi_send_byte(LCD_twi_addr, 0x41); //LCD on

twi_send_byte(LCD_twi_addr, 0xFE);
twi_send_byte(LCD_twi_addr, 0x46); //cursor to home

twi_send_byte(LCD_twi_addr, 0xFE);
twi_send_byte(LCD_twi_addr, 0x52);
twi_send_byte(LCD_twi_addr,   25); //set contrast


twi_send_byte(LCD_twi_addr, 0xFE);
twi_send_byte(LCD_twi_addr, 0x53);
twi_send_byte(LCD_twi_addr,    4); //set backlight
}


//****************************************************************
//void lcd_print_string(char str[])
//****************************************************************

void lcd_print_string(char str[])
{
int k = 0;

while(str[k] != 0x00)
```

```
  {
  twi_send_byte(LCD_twi_addr, str[k]);
  k = k+1;
  delay_10ms();
  }
}


//*****************************************************************
//void move_LCD_cursor(unsigned char position)
//*****************************************************************

void move_LCD_cursor(unsigned char position)
{
twi_send_byte(LCD_twi_addr, 0xFE);
twi_send_byte(LCD_twi_addr, 0x45);
twi_send_byte(LCD_twi_addr, position);
delay_10ms();
}


//*****************************************************************
//void twi_initialize(void)
//*****************************************************************

void twi_initialize(void)
{
//set twi frequency to 100 kHz with ceramic
//resonator frequency at 2.0 MHz
//twi pre-scalar = 1

TWSR = 0;                //no pre-scale
TWBR = ((ceramic_res_freq/scl_clock)-16)/2;
TWCR = TWCR | 0x04;    //TWEN = 1
}


//*****************************************************************
//void twi_send_byte(unsigned char slave_device_addr,
//                     unsigned char send_data);
//*****************************************************************
```

```
void twi_send_byte(unsigned char slave_device_addr,
                   unsigned char send_data)
{
//Send START condition
TWCR = (1<<TWINT)|(1<<TWSTA)|(1<<TWEN);

//Wait for TWINT Flag set. This indicates that the START
//condition has been transmitted
while (!(TWCR & (1<<TWINT)));

//Check value of TWI Status Register. Mask prescaler bits.
//If status different from START go to ERROR
if ((TWSR & 0xF8) != START)
    ERROR(1);

//Load SLA_W into TWDR Register. Clear TWINT bit in
//TWCR to start transmission of address
TWDR = slave_device_addr;
TWCR = (1<<TWINT) | (1<<TWEN);

//Wait for TWINT Flag set. This indicates that the SLA+W has
//been transmitted, and ACK/NACK has been received.
while (!(TWCR & (1<<TWINT)));

//Check value of TWI Status Register. Mask prescaler bits.
//If status different from MT_SLA_ACK go to ERROR
if((TWSR & 0xF8) != MT_SLA_ACK)
    ERROR(2);

//Load DATA into TWDR Register. Clear TWINT bit in TWCR
//to start transmission of data
TWDR = send_data;
TWCR = (1<<TWINT) | (1<<TWEN);

//Wait for TWINT Flag set. This indicates that the
//DATA has been transmitted, and ACK/NACK has been received.
while (!(TWCR & (1<<TWINT)));
```

```
//Check value of TWI Status Register. Mask prescaler bits.
//If status different from MT_DATA_ACK go to ERROR
if((TWSR & 0xF8) != MT_DATA_ACK)
   ERROR(3);

//Transmit STOP condition
TWCR = (1<<TWINT)|(1<<TWEN) | (1<<TWSTO);
}


//***************************************************************
//InitADC: initialize analog-to-digital converter
//***************************************************************

void InitADC( void)
{
ADMUX = 0;                         //Select channel 0
ADCSRA = 0xC3;                     //Enable ADC & start 1st
                                   //dummy conversion
                                   //Set ADC module prescalar
                                   //to 8 critical for
                                   //accurate ADC results
while (!(ADCSRA & 0x10));          //Check if conversion ready
ADCSRA |= 0x10;                    //Clear conv rdy flag -
                                   //set the bit

}


//***************************************************************
//ReadADC: read analog voltage from analog-to-digital converter -
//the desired channel for conversion is passed in as an unsigned
//character variable. The result is returned as a left justified,
//10 bit binary result. The ADC prescalar must be set to 8 to
//slow down the ADC clock at higher external clock frequencies
//(10 MHz) to obtain accurate results.
//***************************************************************

unsigned int ReadADC(unsigned char channel)
{
unsigned int binary_weighted_voltage, binary_weighted_voltage_low;
```

```
unsigned int binary_weighted_voltage_high; //weighted binary
                                  //voltage
ADMUX = channel;                  //Select channel
ADCSRA |= 0x43;                   //Start conversion
                                  //Set ADC module prescalar
                                  //to 8 critical for
                                  //accurate ADC results
while (!(ADCSRA & 0x10));         //Check if conversion ready
ADCSRA |= 0x10;                   //Clear Conv rdy flag - set
                                  //the bit
binary_weighted_voltage_low = ADCL; //Read 8 low bits first
                                  //(important)
                                  //Read 2 high bits,
                                  //multiply by 256
binary_weighted_voltage_high = ((unsigned int)(ADCH << 8));
binary_weighted_voltage = binary_weighted_voltage_low |
                          binary_weighted_voltage_high;
return binary_weighted_voltage;   //ADCH:ADCL
}


//*****************************************************************


void temperatureToLCD(unsigned int ADCValue)

{
float voltage,temperature;
unsigned int tens, ones, tenths;

voltage = (float)ADCValue*5.0/1024.0;
temperature = voltage*100;

tens = (unsigned int)(temperature/10);
ones = (unsigned int)(temperature-(float)tens*10);
tenths = (unsigned int)(((temperature-(float)tens*10)
                                -(float)ones)*10);

twi_send_byte(LCD_twi_addr, ((unsigned char)(tens)+48) );
twi_send_byte(LCD_twi_addr, ((unsigned char)(ones)+48));
```

```
twi_send_byte(LCD_twi_addr, '.');
twi_send_byte(LCD_twi_addr, ((unsigned char)(tenths)+48));
twi_send_byte(LCD_twi_addr, 'F');


}


//*****************************************************************
//delay_10ms: inaccurate, yet simple method of creating delay
// - processor clock: ceramic resonator at 2 MHz
// - nop requires 1 clock cycle to execute
// - 10 ms delay requires 20,000 clock cycles
//*****************************************************************
//*****************************************************************

void delay_10ms(void)
{
unsigned int i,j;

for(i=0; i < 20; i++)
  {
  for(j=0; j < 1000; j++)
    {
    asm("nop"); //inline assembly
    } //nop: no operation
  } //requires 1 clock cycle
}


//*****************************************************************
//delay_10ms: inaccurate, yet simple method of creating delay
// - processor clock: ceramic resonator at 2 MHz
// - nop requires 1 clock cycle to execute
// - 100 ms delay requires 200,000 clock cycles
//*****************************************************************

void delay_100ms(void)
{
unsigned int i,j;
```

```
for(i=0; i < 200; i++)
  {
  for(j=0; j < 1000; j++)
    {
    asm("nop"); //inline assembly
    } //nop: no operation
  } //requires 1 clock cycle
}


//****************************************************************
```

3.8   RADIO FREQUENCY (RF) COMMUNICATIONS THEORY

In the next several sections we discuss Bluetooth and Zigbee communication concepts. These techniques provide a method for microcontrollers to communicate and exchange information via RF communication techniques. In this section we provide a brief tutorial on RF communication concepts.

A fundamental equation in RF communications is the link between a signal's wavelength and frequency of oscillation. The equation, $\lambda = c/v$, indicates the wavelength (λ) is equal to the speed of light (3.0×10^8 m/s) divided by the signal's frequency of oscillation (v). A related concept from antenna theory indicates that to efficiently transfer information from one location to another, the antenna on the transmitter and receiver should ideally be equivalent to the transmitted signal's wavelength. Generally speaking, the higher the frequency of transmission, the smaller the required antenna [USAFA].

Typically, the information we would like to transmit from one location to another is low-frequency (baseband) information as shown in Figure 3.17 as f_i. The baseband is quite crowded with multiple devices trying to exchange this information. To alleviate this situation, the baseband data is modulated or mixed to a higher frequency using a high-frequency carrier signal (f_c). Different RF links may use different carrier signal frequencies and hence be placed at different locations on the RF spectrum [USAFA].

The fundamental component of a modulator is the function multiplier or mixer. The two input signals into the mixer are the low-frequency information signal (f_i) and a high-frequency carrier signal (f_c). The output from the mixer is the sum and difference of the two frequencies. It is important to note the information is now modulated onto the high-frequency carrier and hence smaller antennas may be used for communication.

There are several common methods of modulating the high-frequency carrier signal characteristics with the low-frequency information signal including: amplitude modulation, fre-

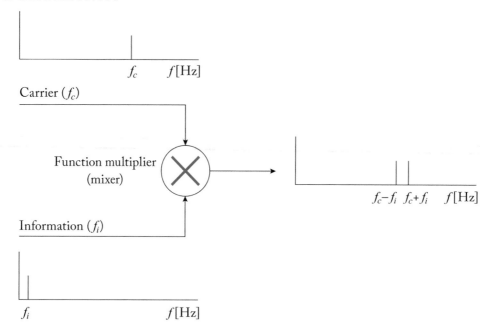

Figure 3.17: Modulation concepts [USAFA].

quency modulation, and phase modulation. As their names imply each modulation technique has a corresponding effect on the high frequency carrier signal parameters.

At the receiver, the modulation process is reversed, called demodulation, to retrieve the desired low-frequency, baseband data.

3.9   BLUETOOTH

Bluetooth (BT) provides for short (maximum 10 m) range RF connections to replace wires. It uses the crowded Industrial, Scientific, and Medical (ISM) frequency band from 2.40–2.50 GHz as shown in Figure 3.18a). BT partitions the band into 79 different, 1 MHz channels (Colbach [8], R&S [16]).

BT employs an interesting frequency-hopping technique to communicate within a configured BT piconet. A piconet consists of a master designated device which provides the hopping sequence to other devices in the piconet. Data for transmission is divided into packets. The master device then transmits a packet of data at the first carrier frequency. It then hops to a different carrier frequency for the next packet and so on until the entire message is transmitted as shown in Figure 3.18b. The transmitter stays at a given carrier frequency for 625 μs which equates to 1,600 hops per second. The master hop sequence is 2^{27} and does not repeat for some time (>23 hours) (Colbach [8], R&S [16]).

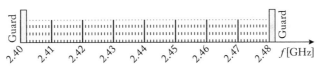

(a) Industrial, scientific, medical (ISM) frequency band

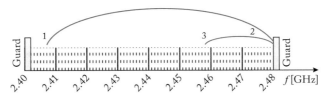

(b) Frequency-hopping spread spectrum (FHSS) [R and S]

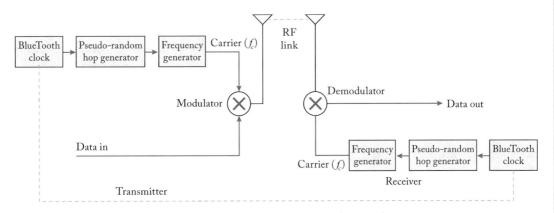

(c) Bluetooth FHSS block diagram [R and S]

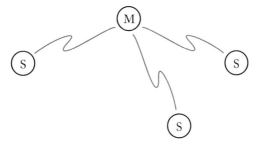

(d) Bluetooth piconet

Figure 3.18: Bluetooth communication concepts (R&S [16]).

Adafruit Bluefruit UART friend   Adafruit Bluefruit SPI friend

Figure 3.19: Adafruit Bluetooth friends [www.adafruit.com].

Formally, the BT modulation technique is called Frequency-Hopping Spread Spectrum (FHSS). A block diagram of the FHSS transmission link is provided in Figure 3.18c). The data for transmission is modulated with the high-frequency hopping carrier. At the receiver the demodulator reverses the process to extract the data. The FHSS transmission technique allows BT to minimize interference in the crowded ISM band (R&S [16]).

BT provides two different types of links: an asynchronous connectionless link (ACL) for data and a synchronous connection oriented (SCO) link for audio and voice applications. The ACL data rate is 432.6 Kbps in a bi-directional, symmetric link while SCO provides a data rate of 64 Kbps (Colbach [8]).

3.9.1   BT HARDWARE AND COMMUNICATIONS

BT communication hardware is available from a number of manufacturers. To provide a microcontroller with BT capability, Adafruit provides a Bluefruit friend that uses the SPI system or UART to connect the BT subsystem as shown in Figure 3.19.

Example 3.1   In this example we connect the Adafruit Bluefruit LE SPI friend with an iPhone. Information is exchanged between the friend and the iPhone. When the letter "l" is sent from the iPhone, an LED connected to pin 5 of the Arduino UNO R3 illuminates for 5 seconds. This demonstrates the capability to send commands from the iPhone and control pins on the UNO R3 via a wireless Bluetooth connection, as shown in Figure 3.20.

The procedures provided here are adapted from *Introducing the Adafruit Bluefruit LE SPI Friend* (Townsend [18])[6] The procedures follow these steps:

---

[6]Please read the entire document on Bluetooth before proceeding.

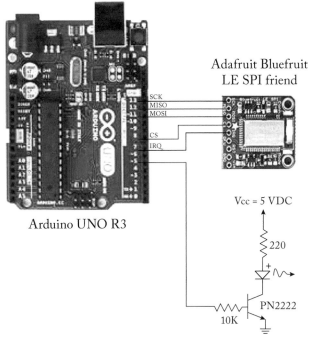 

Arduino UNO R3

Adafruit Bluefruit
LE SPI friend

SCK
MISO
MOSI

CS
IRQ

Vcc = 5 VDC

220

PN2222

10K

LED interface circuit

Adafruit IOS application

Figure 3.20: **Bluetooth lab overview** [www.adafruit.com].

- connections between the Arduino UNO R3 and the Bluefruit LE SPI friend,

- download of the Adafruit Bluefruit LE nRF51 library for the Arduino IDE,

- download of the IOS cell phone application,

- edits to the "bleuart_cmdmode" sketch, and

- testing.

We discuss each step in turn.

Connections.   The connections between the Arduino UNO R3 and the Bluefruit LE SPI friend breakout board are shown in Figure 3.20. An external LED interface circuit is connected to digital pin 5 on the Arduino UNO R3.

Library download.   Download the Adafruit Bluefruit LE Library nRF51. The library is accessible via the Arduino IDE under Tools − > Manage Libraries. The library contains multiple sketches to provide for interaction between the Arduino UNO R3 equipped with the Bluefruit LE SPI friend breakout board and an iPhone.[7]

IOS application download.   The IOS application to interact with Bluefruit devices is available for download at https://adafru.it/f4H.[8]

Sketch edits.   In this example we use the library sketch "bleuart_datamode" to provide two-way communication between the Arduino UNO R3 equipped with the Bluefruit LE SPI friend breakout board and an iPhone. In file "BluefruitConfig.h" the assignments for a hardware-based SPI are retained while all others are commented out or removed. The results are shown in the code below.

```
//************************************************************
//Extract from: BluefruitConfig.h
//
//Adafruit invests time and resources providing this open source code,
//please support Adafruit and open-source hardware by purchasing
//products from Adafruit!
//
//MIT license, check LICENSE for more information
//************************************************************
//COMMON SETTINGS: settings used in SW UART, HW UART and SPI mode
//************************************************************

#define BUFSIZE        128    //Size of read buffer for incoming data
#define VERBOSE_MODE   true   //If set to 'true' enables debug output


//************************************************************
//SHARED SPI SETTINGS: following macros declare the pins to use for
//HW SPI communication.
// - SCK, MISO and MOSI should be connected to HW SPI pins on
//   the Arduino UNO when using HW SPI.
//************************************************************

#define BLUEFRUIT_SPI_CS          8
#define BLUEFRUIT_SPI_IRQ         7
#define BLUEFRUIT_SPI_RST         4
```

---

[7]The library also contains support for the Bluefruit LE UART friend breakout board and also Android devices.
[8]There is also an application available for Android devices.

```
#define BLUEFRUIT_SPI_SCK               13
#define BLUEFRUIT_SPI_MISO              12
#define BLUEFRUIT_SPI_MOSI              11


//*****************************************************************
```

Within the sketch "bleart_datamode," the following changes are required to incorporate the external LED at digital pin 5. In the "void setup(void)" function add the following line as the last line in the function:

```
pinmode(5, OUTPUT);
```

Within the "void loop(void)" function after line:

```
int c = ble.read();
```

add the following code snippet:

```
//*****************************************************************
if(c==0x6C)                 //check for "l" to turn on LED
  {
    digitalWrite(5, HIGH);   //turn the LED on
    delay(5000);
  }
else
  {
    digitalWrite(5, LOW);    //turn the LED off
  }

//*****************************************************************
```

Testing.   Establish connection between the iPhone and the Arduino UNO R3 equipped with the Bluefruit LE SPI. The UART function of the IOS application is used. When an "l" is sent from the IOS application, the LED connected to pin 5 of the Arduino UNO R3 illuminates for 5 seconds.

Extensions.   With the basic connection in place, complete the following activities.

- Connect a red, yellow, and green LED to Arduino digital pins 3, 4, and 5, respectively. Illuminate each LED when a "r", "y", or "g" character is sent from the IOS application.

- Connect an LM34 Precision Fahrenheit Temperature Sensor to Arduino analog input A0. Send the sensed temperature to the Iphone for display.

3.10   ZIGBEE

Zigbee technology provides for a network of low-power, battery-operated nodes connected via RF links. Zigbee has become popular in assembling networks for applications such as home security and smart home applications (Zigbee [20]).

A Zigbee network consists of nodes of three different types and capabilities (Eady [9]).

- Zigbee Coordinator (ZC): The ZC as its name implies organizes nodes into ad hoc networks. There is one ZC per network. It contains the network security key and serves as the root node for the network. It has the capability to make connections to other networks.

- Zigbee Router (ZR): The ZR provides links to another nodes and passes data to the other nodes.

- Zigbee End Device (ZED): The ZED cannot link to other ZED configured nodes; however, it has the important function of collecting data and providing control signals to end devices such as sensors and actuators.

Once a network is formed of Zigbee configured devices as shown in Figure 3.21a, it has the capability to share secure data over a wide area at a data rate of up to 250 kit per second. An individual Zigbee configured device may have a range up to 100 m indoors and up to 300 m outdoors in a line of site configuration. Packet encryption is provided via the AES-128 encryption standard (Eady [9]).

As shown in Figure 3.21b,c, Zigbee uses the crowded Industrial, Scientific, and Medical (ISM) frequency band from 2.40–2.50 GHz. Each Zigbee channel (11–26) is 2 MHz wide with 5 MHz spacing between adjacent channels [metageek.com].

A number of manufacturers provide Zigbee and XBEE-compatible hardware. XBEE is a line of compatible devices sharing a common interface and form factor. XBEE devices support Zigbee as well as other protocols. Figure 3.21d shows the components of an XBEE experimenter kit from Sparkfun [sparkfun.com]. We set up an XBEE network using the kit in the next section.

3.10.1   SPARKFUN XBee3 WIRELESS KIT

The components of the Sparkfun XBee3 Wireless Kit[9] are shown in Figure 3.22a. The kit contains an XBEE shield, an XBEE Explorer, and two XBEE3 modules. To assemble a two-chat between the XBEE3 modules, an XBEE network is assembled as shown in Figure 3.22b.

---

[9]Sparkfun provides an excellent kit tutorial titled "XBEE Shield Hookup Guide." Please read this guide before proceeding.

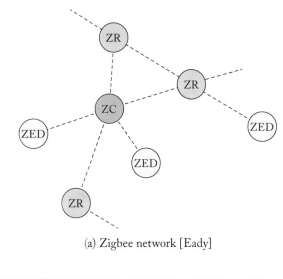

(a) Zigbee network [Eady]

(b) Industrial, scientific, medical (ISM) frequency band

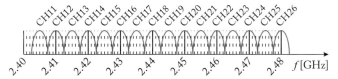

(c) Zigbee channels with the ISM spectrum [metageek.com]

(d) XBEE wireless kit [sparkfun.com]

Figure 3.21: Zigbee networks. Images used courtesy of Sparkfun (CC BY 2.0) [www.sparkfun.com].

XBEE shield

XBEE Explorer

XBEE3 modules
with PCB antennas

(a) Sparkfun XBEE KIT-15936

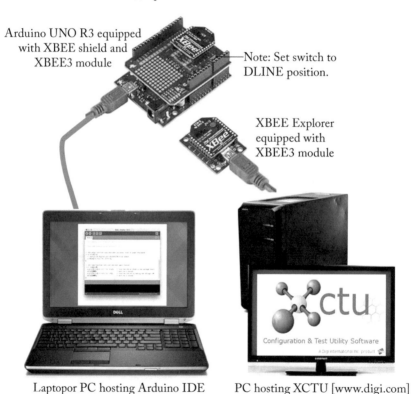

Arduino UNO R3 equipped
with XBEE shield and
XBEE3 module

Note: Set switch to
DLINE position.

XBEE Explorer
equipped with
XBEE3 module

Laptopor PC hosting Arduino IDE

PC hosting XCTU [www.digi.com]

(b) XBEE network configuration

Figure 3.22: **XBEE networks.** Images used courtesy of Sparkfun (CC BY 2.0) [www.sparkfun.com].

The XBEE shield is mounted to an Arduino UNO R3. Ensure the DLINE/UART switch is set to the DLINE position. The host PC/laptop for the shield should be equipped with the Arduino IDE. The XBEE Explorer equipped with an XBEE3 module is hosted by another PC/laptop equipped with the DIGI XCTU Configuration and Test Utility Software. This is available for download from DIGI [www.digi.com]. You also need to download and install an FT231X driver from a reliable source. I downloaded the driver from Dell. Using the XCTU software both of the XBEE3 modules should be updated with the latest 802.15.4 firmware.

Within the "XBEE Shield Hookup Guide" is an Arduino sketch titled "XBEE_serial_passthrough.ino." The sketch should be compiled and uploaded to the XBEE shield using the Arduino IDE.

The XCTU software is then used to configure both devices into an XBEE network. Switch the XCTU to Console mode, close the serial connection with the radio module, and commence chatting.

Extension. The "XBEE Shield Hookup Guide" also contains an Arduino sketch titled "XBee_Remote_Control.ino." Use the sketch to remotely control some feature on the Arduino shield via the XCTU software.

3.11   CELLULAR MICROCONTROLLER COMMUNICATIONS

Much of the planet is accessible via the cellular communications network. It is important to note the cellular communications network is continually evolving and is not homogenous throughout the globe. The first digital based network was designated 2G for second generation. The digital cellular system has evolved from 2G to 3G to 4G (or LTE for Long Term Evolution) to now 5G. As you plan a cellular-based IoT application, it is important to carefully examine the generation of network capability for the deployed product and the long-term future of the corresponding technology. As an example, much of the early IoT cellular applications were based on 2G technology. However, the long-term viability of 2G technology is in doubt in different portions of the globe (Remmert [17]).

A cellular-based IoT application typically requires a Subscriber Identity Module (SIM) card. The SIM card is removable and contains subscription information and the user's phone book [adafruit.com]. The SIM card is activated through a cellular network provider that also will provide a subscription to their cellular services. It is important to emphasize in choosing a cellular shield or support card, an appropriate SIM and subscription service must be available to support the underlying technology where the product will be deployed. In the next several sections we discuss a 2G GSM-based shield and then a 4G LTE-based shield. Both types of shields are reviewed to support our readers' needs across the globe.

3.11.1   ADAFRUIT 2G FONA 800

Adafruit provides a shield to allow an Arduino-based processor to send and receive information (voice and data) via the GSM 2G network.[10]

To connect an Arduino processor to the cellular network requires [www.adafruit.com]:

- an Arduino processor,

- an Adafruit FONA 800 shield (#2468) or breakout board,[11]

- a microphone (#1064) and speaker (#1890) or a cell phone compatible TRRS headset (#1966),

- a FONA 800 compatible antenna (#1991),

- a 3.7 VDC, 1200 mAH Li-Po battery (#258), and

- a 2G SIM card (#2502).

The FONA 800 components are conveniently available through Adafruit as a starter pack [adafruit.com].

The Adafruit 800 is a powerful shield that allows access to FM radio, the Global Positioning System (GPS), and the worldwide cellular network. In this section we examine some of these features. To get started download the Arduino library for the Adafruit FONA. It is available via the Arduino IDE via Tools − > Manage Libraries.

Once the library has been downloaded, activate the SIM card you have chosen to use with the FONA 800.

FM Testing.   Note: This activity does not require a SIM card. Upload the sketch FONAtest from the FONA 800 library to the Arduino UNO R3 equipped with the FONA 800 system shown in Figure 3.23. When the serial monitor is opened, a menu of activities is printed. To receive a local FM station, use the following steps.

- H: set headphone audio.

- M: get FM volume.

- m: set FM volume.

- f: tune FM radio (Set your favorite station frequency as a 3–4 digit code, e.g., 91.9 MHz is set as 919).

- Enjoy your station!

After completing the FM activity, complete GPS testing and cell phone testing using sketches available in the FONA 800 library. These activities require an activated SIM card.

---

[10]The 2G network is being phased out by many cellular providers.

[11]Please ensure these components are compatible with your location on the planet.

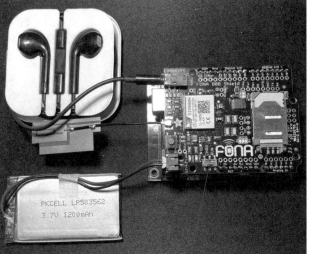

Cell-phone TRRS headset #1966

GSM/cellular quad-band antenna #1991

Lithium Ion Polymer (LiPo) battery – 3.7 V, 1200 mAh #258

FONA 800 shield voice/data cellular GSM #2468

GSM SIM card from Ting & Adafruit #2505

Arduino UNO R3 (under shield)

Charge/run switch

Figure 3.23: Adafruit FONA 800 [adafruit.com].

3.11.2   SPARKFUN LTE SHIELD

Sparkfun provides a LTE CAT M1/NB-IoT shield (CEL-15087). The shield is hosted on an Arduino-based processor such as the Arduino UNO R3, as shown in Figure 3.24. Sketches loaded onto the UNO R3, sends commands and receive information from the LTE shield. The shield is equipped with an onboard antenna or may be equipped with an external antenna via a U.FL connector. An LTE-compatible "pay-as-you-go" SIM card and subscription service is available from Hologram and may be purchased via Sparkfun [www.sparkfun.com]. I found the Hologram easy to use and activate.

Sparkfun provides a step-by-step tutorial "LTE Cat M1/NB-IoT Shield Hookup Guide" by Jim Blom [www.sparkfun.com]. A thorough review of this excellent tutorial is recommended before proceeding. To get the cellular shield up and operating the following steps are completed.

- Solder header pins to the shield.

- Mount the assembled shield to the UNO R3.

- Activate the Hologram SIM card at https://dashboard.hologram.io/account/login. You will need to establish an account and provide credit card information. You are charged for the services used.

- Once the card is activated insert it into LTE shield.

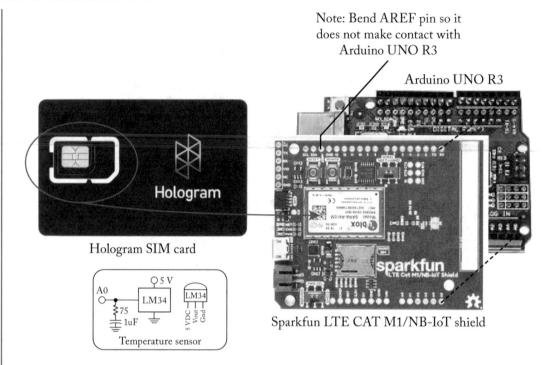

Note: Bend AREF pin so it
does not make contact with
Arduino UNO R3

Arduino UNO R3

Hologram SIM card

Sparkfun LTE CAT M1/NB-IoT shield

Figure 3.24: Sparkfun LTE CAT M1/NB-IoT shield. Images used courtesy of Sparkfun (CC BY 2.0) [www.sparkfun.com].

- Ensure the switch settings on the shield are set for "Arduino" on the "PWR_SEL" switch and "SOFT" on the "SERIAL" switch.

- Connect the Arduino with the LTE shield attached to the host PC/laptop.

- Start up the Arduino IDE and download the "Sparkfun LTE Shield Arduino Library" using the Arduino IDE Library Manager.

- With the library downloaded, start up the "00_Register_Operator" sketch and upload it to the Arduino UNO R3.

- I used the "DEFAULT" setting within the sketch. The sketch will scan for available cellular providers in the area and provide a list of Mobile Country Code (MCC) and Mobile Network Code (MNC) combinations. The list is provided on the serial monitor. You can identify the specific provider by looking up the MCC/MNC combination on the Roaming Zone website [www.roamingzone.com].

- Select the desired provider from the list. The shield will then establish connection with the provider. Connection will be indicated by the blue "NET" LED on the LTE shield.

- With NET connection established, load the "01_SMS_SEND" sketch to send a text from the LTE shield to a cell phone.

After a text has successfully been sent, connect an LM34 Precision Fahrenheit Temperature Sensor to the A0 pin of the Arduino UNO R3. **Note:** Sparkfun indicated the AREF pin was accidently connected to ground during manufacturer of the LTE shield. The AREF pin on the shield should be bent out of the way so it does not connect with the Arduino UNO R3.

Modify the "01_SMS_SEND" sketch with the code shown below. This snippet will text the LM34 sensed temperature to a cell phone. The code should be inserted into the sketch after:

```
message = ""; // Clear message string
```

code to insert:

```
//send temp
int A0;
A0 = analogRead(A0);
A0 = map(A0, 1,1024, 0, 500);    //remap ADC A0 value to temp
messageToSend = "Greenhouse temp: " + String(A0); // Add A0 to
messageToSend += "\r\n"; // Create a new line
//messageToSend += "Time = " + String(time);
lte.sendSMS(DESTINATION_NUMBER, messageToSend);
```

3.12 APPLICATION: NEAR FIELD COMMUNICATIONS (NFC)

Earlier in the chapter we discussed NFC communications. Adafruit manufacturers several products including a shield and breakout board to explore NFC concepts. The boards may be used with a variety of Arduino-based products via a SPI or I2C connection. In this example we connect an Arduino UNO R3 to the Adafruit NFC + RFID Shield (#789). This example and accompanying code was adapted from "Adafruit PN532 RFID/NFC Breakout and Shield." A thorough review of this excellent tutorial is recommended before proceeding.

To provide for NFC communications, the following steps are required.

- Solder the header pins to the Adafruit NFC Shield.

- Download the Adafruit PN532 software library.[12]

- Adapt the Adafruit readMiFare sketch for I2C operation.

---

[12]The library is accessible via the Arduino IDE under Tools − > Manage Libraries.

 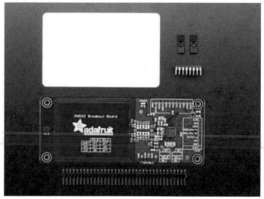

PN532 NFC/RFID Controller Shield for Arduino (#789) and PN532 NFC/RFID controller breakout board (#364)

Figure 3.25: Adafruit NFC products [www.adafruit.com].

The Adafruit readMiFare sketch contains code for both SPI and I2C operation. It also provides a number of activities for interaction between the NFC shield and the 13.56 MHz RFID card (Adafruit #359). I also purchased a 13.56 MHz keychain fob. The readMiFare sketch was adapted to distinguish between different cards based on the card/fob ID. This allows specific actions to be initiated (e.g., unlock a door) when a specific card or fob is present. The UML activity diagram for the sketch is provided in Figure 3.26.

Once the sketch is verified and uploaded, the status of nearby NFC cards and fobs may be read within the Adafruit IDE serial monitor window. Ensure the BAUD rate is set for 115,200 within the serial monitor window.

```
//************************************************************
//This example was adapted from Adafruit sketch readMifare
//@file     readMifare.pde
//@author   Adafruit Industries
//@license  BSD (see license.txt)
//
//This example will wait for any ISO14443A card or tag with a four byte
//UID size.  These include:
// - Adafruit 13.56 MHz classis keychain fob #363
// - Adafruit 13.56 MHz RFID card #359
//
//When one of these devices is within range of the PN532 RFID/NFC
//shield, the sketch:
```

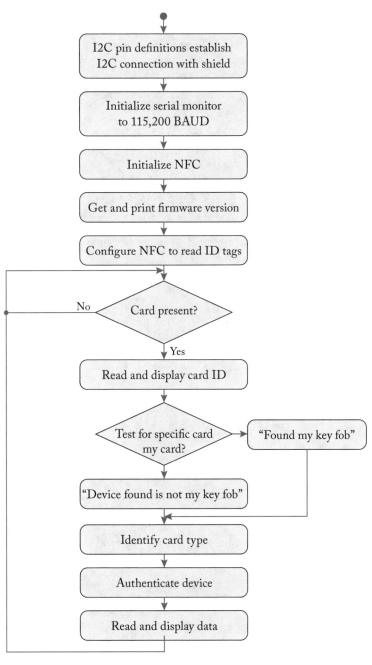

Figure 3.26: **NFC UML for modified readMiFare sketch** [www.adafruit.com].

```
//    - Displays card information
//    - Tests for a specific device
//    - Authenticate block 4 (the first block of Sector 1) using
//      the default KEYA of OXFF OXFF OXFF OXFF OXFF OXFF
//    - If authentication succeeds, we can then read any of the
//      4 blocks in that sector (though only block 4 is read here)
//
//Adafruit invests time and resources providing this open source code,
//please support Adafruit and open-source hardware by purchasing
//products from Adafruit!
//
//Modified by: S. Barrett, Dec 16, 2020
//   - Added test for specific device
//***********************************************************************

#include <Wire.h>
#include <Adafruit_PN532.h>

//I2C pin definitions for IRQ and reset lines
#define PN532_IRQ    (2)
#define PN532_RESET (3)

//I2C connection:
Adafruit_PN532 nfc(PN532_IRQ, PN532_RESET);

void setup(void)
{
Serial.begin(115200);
Serial.println("Hello!");
nfc.begin();
uint32_t versiondata = nfc.getFirmwareVersion();
if(! versiondata)
  {
  Serial.print("Didn't find PN53x board");
  while (1); // halt
  }

// Got ok data, print it out!
```

```
Serial.print("Found chip PN5");
Serial.println((versiondata>>24) & 0xFF, HEX);
Serial.print("Firmware ver. ");
Serial.print((versiondata>>16) & 0xFF, DEC);
Serial.print('.');
Serial.println((versiondata>>8) & 0xFF, DEC);

//configure board to read RFID tags
nfc.SAMConfig();

Serial.println("Waiting for an ISO14443A Card ...");
}


void loop(void)
{
int j, fob_found;
uint8_t success;
uint8_t uid[] = { 0, 0, 0, 0, 0, 0, 0 };  //Buffer for returned UID
uint8_t fob[] = {0xB5, 0xE1, 0x56, 0xBC};
uint8_t uidLength;                        //Length of the UID - 4 bytes

//Wait for an ISO14443A type cards (Mifare, etc.).  When one is found
//'uid' will be populated with the UID, and uidLength.
//Sketch has been modified for 4 byte uidLengthwill
success=nfc.readPassiveTargetID(PN532_MIFARE_ISO14443A, uid,
                                &uidLength);
if(success)
  {
  //Display some basic information about the card
  Serial.println("Found an ISO14443A card");
  Serial.print("  UID Length: ");
  Serial.print(uidLength, DEC);
  Serial.println(" bytes");
  Serial.print("  UID Value: ");
  nfc.PrintHex(uid, uidLength);
  Serial.println("");
```

```
   //Test for specific device
   fob_found = 1;
   for(j=0; j<4; j++)
     {
     if((uid[j] == fob[j])&&(fob_found == 1))
       fob_found = 1;
     else
       fob_found = 0;
     }
if(fob_found == 1)
  Serial.println("Found my keyfob");
else
  Serial.println("Device found is not my keyfob");


 if(uidLength == 4)
    {
    //We probably have a Mifare Classic card ...
    Serial.println("Seems to be a Mifare Classic card (4 byte UID)");

    //Now we need to try to authenticate it for read/write access
    //Try with the factory default KeyA: 0xFF 0xFF 0xFF 0xFF 0xFF 0xFF
    Serial.println("Authenticating block 4 with default KEYA value");
    uint8_t keya[6] = { 0xFF, 0xFF, 0xFF, 0xFF, 0xFF, 0xFF };

    //Start with block 4 (the first block of sector 1) since sector 0
    //contains the manufacturer data and it's probably better just
    //to leave it alone unless you know what you're doing
    success=nfc.mifareclassic_AuthenticateBlock(uid,uidLength,4,
                                      0,keya);
    if (success)
      {
      Serial.println("Sector 1 (Blocks 4..7) has been authenticated");
      uint8_t data[16];

      //Try to read the contents of block 4
      success = nfc.mifareclassic_ReadDataBlock(4, data);
```

```
        if(success)
          {
          //Data seems to have been read ... spit it out
          Serial.println("Reading Block 4:");
          nfc.PrintHexChar(data, 16);
          Serial.println("");
          //Wait a bit before reading the card again
          delay(1000);
          }
        else
          {
          Serial.println("Ooops ... unable to read the requested"
                      "block.");
          Serial.println("Try another key?");
          }
        }
    else
      {
      Serial.println("Ooops...authentication failed: Try another "
                  "key?");
      }
  }//uidLength == 4
 }
 }


//****************************************************************
```

3.13 SUMMARY

The goal of this chapter was to provide a short introduction of methods of connecting a micro-controller to external peripheral devices and other microcontrollers. The chapter began with a brief description of serial communications and related serial communication terminology. We then reviewed different connectivity technologies beginning with close range technologies, then mid-range technologies, and concluded with RF technologies.

3.14 REFERENCES

[1] Ada, L. *Adafruit PN532 RFID/NFC Breakout and Shield.* www.adafruit.com

[2] Ada, L. *Adafruit FONA 800 Shield.* www.adafruit.com

[3] Arduino homepage. www.arduino.cc

[4] Barrett, S. and Pack, D. (2006). *Microcontrollers Fundamentals for Engineers and Scientists*, Morgan & Claypool Publishers. DOI: 10.2200/s00025ed1v01y200605dcs001.

[5] Barrett, S. and Pack, D. (2019). *Atmel AVR Microcontroller Primer Programming and Interfacing*, 3rd ed., Morgan & Claypool Publishers. DOI: 10.2200/s00100ed1v01y200712dcs015.

[6] Barrett, S. (2020). *Arduino II: Systems*, Morgan & Claypool Publishers. DOI: 10.2200/s01024ed1v01y202006dcs059.

[7] Blom, J. *XBEE Shield Hookup Guide*. www.sparkfun.com 181

[8] Colbach, G. (2019). *Wireless Networking—Introduction to Bluetooth and WiFi*. gordoncolbach@cloudversity.com 106, 108, 130

[9] Eady, F. (2007). *Hands-On ZigBee: Implementing 802.15.4 with Microcontrollers*, Newness. 112

[10] *Electrical Signals and Systems*, Department of Electrical Engineering, United States Air Force Academy, McGraw Hill.

[11] *The I2C—Bus Specification*, Version 2.1, Philips Semiconductor, January 2000. 91

[12] Maxim Integrated (2019). *Fundamentals of NFC/RFID Communication*. 59, 60, 61

[13] *Microchip ATmega328 PB AVR Microcontroller with Core Independent Peripherals and Pico Power Technology DS40001906C*, Microchip Technology Incorporation, 2018. www.microchip.com

[14] Barrett, S. and Pack, D. (2019). *Microchip AVR Microcontroller Primer: Programming and Interfacing*, 3rd ed., Morgan & Claypool Publishers.

[15] Nilsson, J. and Reidel, S. (1996). *Electric Circuits*, 5th ed., Addison Wesley. 60, 61

[16] Roland, R. (2016). *Bluetooth Adaptive Frequency Hopping on a R&S CMW*, Application Note 11.2016-1C108-0e, Rohde & Schwarz. www.rohde-schwarz.com 106, 107, 108

[17] Remmert, H. (2020). *2G, 3G, 4G LTE Network Shutdown Updates*. www.digi.com 115

[18] Townsend, K. *Introducing the Adafruit Bluefruit LE SPI Friend*. www.adafruit.com 108

[19] Triggs, R. *What is NFC and How Does it Work*. androidauthority.com 59, 60, 61

[20] Zigbee Alliance. www.zigbeealliance.org 112

3.15 CHAPTER PROBLEMS

3.1. Summarize the differences between parallel and serial bit stream conversion techniques.

3.2. Summarize the differences between the USART, SPI, and TWI methods of serial communication.

3.3. Draw a block diagram of the USART system, label all key registers, and all keys USART flags.

3.4. Draw a block diagram of the SPI system, label all key registers, and all keys SPI flags.

3.5. Draw a block diagram of the TWI system, label all key registers, and all keys TWI flags.

3.6. If an ATmega328 microcontroller is operating at 12 MHz, what is the maximum transmission rate for the USART and the SPI?

3.7. What is the ASCII encoded value for "ATmega328"?

3.8. Draw the schematic of a system consisting of two ATmega328s that will exchange data via the SPI system.

3.9. Write the code to implement the system described in the question above.

3.10. Modify the TWI example provided in the chapter to read the temperature from a TMP 102 digital temperature sensor and display the result on an LCD. Provide a schematic and program.

3.11. Are there any limitations to connecting multiple devices to a TWI configured network? Explain.

3.12. It is desired to equip the ATmega328 with eight channels of digital-to-analog conversion. What serial communication system should be employed? Provide a detailed design.

3.13. Research the BlinkM Smart LED. Provide a schematic to connect the LED to the ATmega328.

3.14. Write a program to implement the interface between the ATmega328 and the BlinkM Smart LED.

3.15. Research Sparkfun's 6.5-inch seven-segment displays. Provide a schematic to connect the displays to the ATmega328. Write a program to implement the interface between the ATmega328 and the displays.

3.16. Design a table of connectivity features. On the horizontal axis list the following communication techniques: Bluetooth, Zigbee, NFC, GSM, ethernet, I2C, SPI, and USART. On the vertical axis include such parameters as range, data rate, etc. Complete the table.

CHAPTER 4

Application: IoT Greenhouse

Objectives: After reading this chapter, the reader should be able to do the following.

- Apply instrumentation and IoT concepts to design a project.

- Describe tools used to systematically design hardware and software for an instrumentation project.

4.1 OBJECTIVE

The objective of this chapter is to demonstrate in action the concepts discussed in this book. Simply put, our goal is to provide the theory, design, and construction of a passively heated greenhouse. We equip the greenhouse with instrumentation to monitor and control key parameters. Using IoT concepts, key parameters will be monitored and controlled remotely. Via a weather station and greenhouse controller examples, we demonstrate how to interface different sensors and peripheral devices to the Arduino UNO R3 and the MKR 1000. These examples may be used and adapted for many other non-greenhouse systems.

We begin the chapter with the theory of greenhouse design. The reader is assumed to have no background in this area. This section was compiled from a number of excellent sources listed at the end of the chapter. We then present the design and construction of a passively heated greenhouse. In our example an existing 8×8 foot garden shed is converted into a greenhouse. This is a do-it-yourself (DIY) project. Prior to this project, the author had no background in greenhouse design and construction. However, it was a project considered for some time. The chapter then provides details on an Arduino-based greenhouse instrumentation and control system. Finally, key greenhouse parameters are made available for monitoring and control using Arduino-based IoT concepts. Prior to continuing, the reader is encouraged to review microcontroller-based system design concepts provided in *Arduino II: Systems,* Chapter 7.

4.2 ASIDE: LOCAL vs. REMOTE OPERATION

Throughout the book, we assume the ready availability of a support PC/laptop to provide access to the Arduino IDE serial monitor, power for the microcontroller and its accompanying shields, etc. It is important to keep in mind that in developing an IoT application, the support PC/laptop will typically not be available. Furthermore, power for the IoT device may be provided by a battery. The examples provided in this chapter are for a remote IoT application to instrument a greenhouse. Power is provided by a solar panel with a battery backup.

During software development, it is helpful to use Arduino IDE features such as the serial monitor. In a remote application the Arduino IDE will not be available. Rather than develop two versions of your code (local version with Arduino IDE support and a remote version), a global variable may be used to "comment out" code used in development (local) that will not be needed during remote operation.

For example, a global variable "local" may be declared. When set to logic one, indicating local operation, the code is executed. When set to logic zero (indicating remote operation), the code is bypassed. The code is placed in an "if" statement:

```
//*******************************************************************

int   local = 1;            //set to 1 for local, 0 for remote

if (local)
 {
 :
//insert code for local operations only
 :
 }

//*******************************************************************
```

In this technique all the code for local and remote operation is complied and stored within the processor. This may become an issue with larger programs when microcontroller memory assets begin to fill. An alternative approach to conserve memory assets is to used conditional compilation techniques. In this technique a variable is defined and set to determine if included code is compiled (Kelley [8]).

```
//*******************************************************************

#define ECHO_TO_SERIAL  0        //set to 1 for local, 0 for remote

#if ECHO_TO_SERIAL

//insert code for local operations only

#endif

//*******************************************************************
```

4.3 GREENHOUSE THEORY

There is considerable information available to guide the design of a passive greenhouse. A passive greenhouse uses solar energy to either extend the growing season for plants or to grow plants year round. Equally important is to provide for greenhouse cooling and ventilation during hot summer months. The information provided here is a compilation of the excellent sources listed at the end of this chapter. A thorough review of these sources is recommended.

Passive heating uses the energy of the sun to heat the interior of the greenhouse during the day. The heat energy is stored using a large thermal mass such as barrels filled with water. When the temperature drops at night, the energy stored in the water is released into the greenhouse to mitigate internal temperature fluctuations.

The British Thermal Unit (BTU) is the energy required to raise one pound of water a degree Fahrenheit. So during the day, the water barrels absorb energy. At night a drop of one degree per pound of water would release one BTU of heat energy.

To efficiently capture the solar energy, greenhouse windows should face south. Windows may be included on the east and west sides as well. Typically, the north facing wall is not equipped with windows but is thoroughly insulated.

The south facing windows ideally should be inclined at an angle. A general rule of thumb is to take your location's latitude and add ten degrees. I live in Laramie, Wyoming (41.3114° N, 105.5911° W); therefore, ideally the south facing windows should be inclined at 51°.

To determine the amount of water required for passive energy storage several rules of thumb are used.

- To extend the growing season, 2.5 gallons of water needs to be stored for every square foot of glazing material (windows).

- To grow plants year round, 5 gallons of water needs to be stored for every square foot of glazing material.

- It is worth noting that a gallon of water weighs 8.3 pounds.

To maximize the collection of solar energy, it was decided to place windows on the south, east, and west walls of the greenhouse. In this specific example, an 8 × 8 foot by 8 foot existing garden shed was converted to a greenhouse.

A combination of mobile home glass windows and plexiglass bubble (RV skylight) windows were installed. The total glazing area was 36.2 square feet. To provide for year round growing, 180 gallons of water is required for passive storage. A total of 24, five-gallon, black plastic buckets were used. The balance of water storage was accomplished using a 55-gallon rain barrel. The buckets were placed on shelves within the greenhouse. Realize the greenhouse floor will be supporting 180 gallons (1,500 pounds) of water. Ensure the floor is adequately supported to do this. I included some additional 2 inch × 4 inch support under the floor for this purpose.

The shed has traditional 2 inch × 4 inch framing covered by 1/2-inch-thick exterior wallboard. The roof of the shed was peaked and covered with roofing shingles as shown in Figure 4.1 top left. The north-facing wall of the shed is insulated with R-13 insulation.

The shed's interior is insulated and covered with wallboard in non-window locations. The interior roof was also insulated and covered with 1 inch × 4 inch pine planks. The interior was painted white for reflectivity. The shed floor was covered with a 1/4-inch black rubber mat. A 12 VDC exhaust fan was installed in the east facing eave. Finally, windows were covered on the shed interior with 8-inch plexiglass to provide an additional thermal barrier. The greenhouse exterior and interior are shown in Figure 4.1.[1]

4.4   GREENHOUSE INSTRUMENTATION SYSTEM

The Greenhouse Instrumentation System (GIS) has the following requirements:

- self-contained solar power with a 12 VDC battery backup,

- compartmentalized for systematic design and expansion, and

- Arduino-based technology for greenhouse instrumentation and control.

The GIS block diagram is shown in Figure 4.2.

In keeping with a compartmentalized design, the GIS has four subsystems:

- solar power system;

- weather station;

- greenhouse control; and

- Arduino-based IoT interface.

Each subsystem is discussed in turn.

4.5   SOLAR POWER SYSTEM

The solar power systems was discussed in *Arduino I: Getting Started*. An excerpt is included here with permission for completeness.

The solar power system consists of a solar panel, a solar power manager, a rechargeable battery, and fuses for circuit protection. In this project we use the DFRobot DFR0580 Solar Power Manager for a 12 VDC lead-acid battery. With an 18 VDC, 100 W solar panel, and a 12 VDC lead-acid battery, the DFR0580 can provide regulated output voltages of 5 VDC at 5 amps and 12 VDC at 8 amps. This is suitable for the GIS project [www.DFRobot.com]. A diagram of the solar power system is shown in Figure 4.3.

---

[1]Morgan & Claypool is a global publisher. Due to the wide range of applicable codes and standards and permitting requirements, please check with local requirements concerning permitting and installation requirements.

Figure 4.1: **Greenhouse project.**

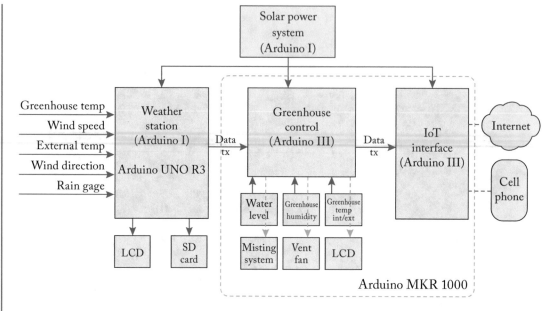

Figure 4.2: Greenhouse instrumentation system.

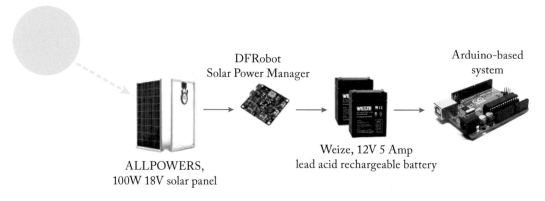

Figure 4.3: Solar power system. Images courtesy of AllPowers, DFRobot, Weize, and Arduino.

A distribution panel was designed for the overall system and is shown in Figure 4.4. The various components are connected together as shown in the figure with an automotive style fuse block and automotive style blade fuses.

The panel is housed in a QILIPSU plastic, hinged 1.6 inch × 10.6 inch × 5.9 inch enclosure.

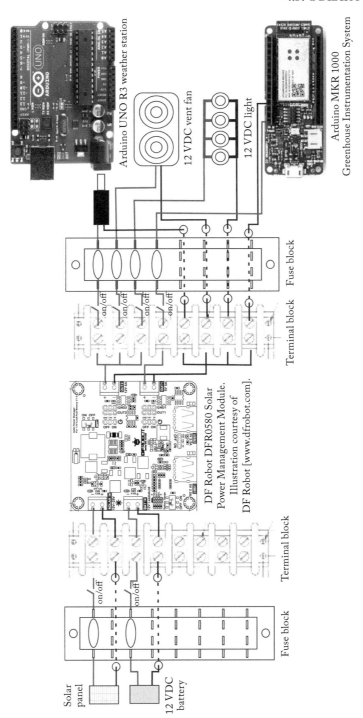

Figure 4.4: Solar power distribution panel. Arduino illustrations used with permission of the Arduino Team (CC BY-NC-SA) [www.arduino.cc].

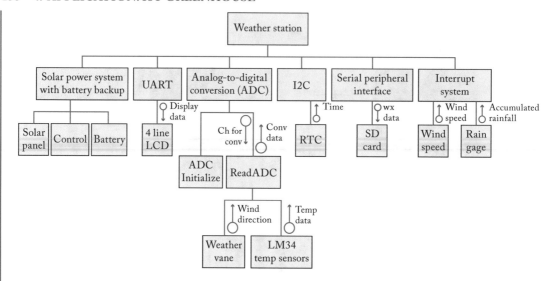

Figure 4.5: Weather station structure chart.

4.6   WEATHER STATION

In this project we design a weather station with the following requirements[2]:

- design a weather station to sense wind direction and speed, ambient temperatures, and accumulated rainfall; and

- collected weather data should be displayed on a two line LCD and transmitted serially and time stamped to a microSD card for storage.

4.6.1   STRUCTURE CHART

To begin, the design process, a structure chart is used to provide an overall system picture and partition the system into definable pieces. The structure chart for the weather station is provided in Figure 4.5. We employ a top-down design/bottom-up implementation approach. The system is partitioned until the lowest level of the structure chart contains "doable" pieces of hardware components or software functions. Data flow is shown on the structure chart as directed arrows.

The main microcontroller subsystems needed for this project are the USART (serial port) for the LCD display, the ADC system for wind direction and temperature sensing, the I2C system to communicate with the Real Time Clock (RTC) module, the Serial Peripheral Interface (SPI) for the SD card data logger, and the interrupt system for the wind speed and the rain gage.

---

[2]This project originally appeared in *Arduino I: Getting Started*, Morgan & Claypool Publishers, 2020. An expanded version is included here with permission as an integral part of the Greenhouse Project.

4.6.2   CIRCUIT DIAGRAM

The circuit diagram for the weather station and subsystems is provided in Figure 4.6. We discuss each subsystem in turn.

Liquid Crystal Display.   We use the Sparkfun LCD-09535 serial enabled, 16-character, 2-line LCD to display weather data. To display four lines of weather data we toggle between two-line displays.

Temperature Sensors.   The weather station is equipped with two LM34 Precision Fahrenheit Temperature Sensors. The LM34 provides 10 mV output per degree Fahrenheit. The output from the sensors are provided to Arduino analog input pin A0 for the external sensor and pin A1 for the internal sensor (SNIS161D).

Weather Vane.   The Sparkfun Weather Meters kit (# SEN-08942) is equipped with a weather vane, an anemometer (wind speed), and a rain gauge. The weather vane provides a voltage output from 0–5 VDC for different wind directions, as shown in Figure 4.6. The weather vane must be oriented to a known direction with the output voltage at this direction noted. The output voltage is provided to an RJ11 connector. Pins 1 and 4 of the RJ11 connector provide access to the vane's resistance. A 10-KOhm resistor is placed in series with the vane's resistance to provide a voltage divider circuit [www.sparkfun.com]. The weather vane output is connected to Arduino analog input A3.

Rain Gauge.   The rain gauge contains an enclosed rain tipping bucket. When the bucket is full, it tips, tripping a magnetic read switch. The switch closure is routed via an RJ11 connector to Arduino interrupt pin 3. Each tip represents 0.011 inch (0.2794 mm) of rainfall.

Anemometer-wind Speed.   The anemometer is equipped with three cups to catch the wind. With each rotation of the anemometer, a reed switch is tripped. Switch closures 1 second apart equates to 1.492 miles per hour (2.4 Km/h) of wind speed. The switch closure is routed via an RJ11 connector to Arduino interrupt pin 2. As an alternative, we also demonstrate a non-interrupt technique to measure wind speed.

SD Card with Real Time Clock.   We use the Adafruit Data Logger Shield (Adafruit #1141). The shield provides multiple gigabits of additional storage for the UNO R3. It features an RTC with battery backup to timestamp collected data. RTC features provide the microcontroller the ability to track calendar time based on seconds, minutes, hours, etc. [www.adafruit.com].

Solar Power System.   The solar power system consists of a solar panel, a solar power manager, a rechargeable battery, and fuses for circuit protection. In this project we use the DFRobot DFR0580 Solar Power Manager for a 12 VDC lead-acid battery. With an 18 VDC, 100W solar panel and a 12 VDC lead-acid battery; the DFR0580 can provide regulated output voltages of

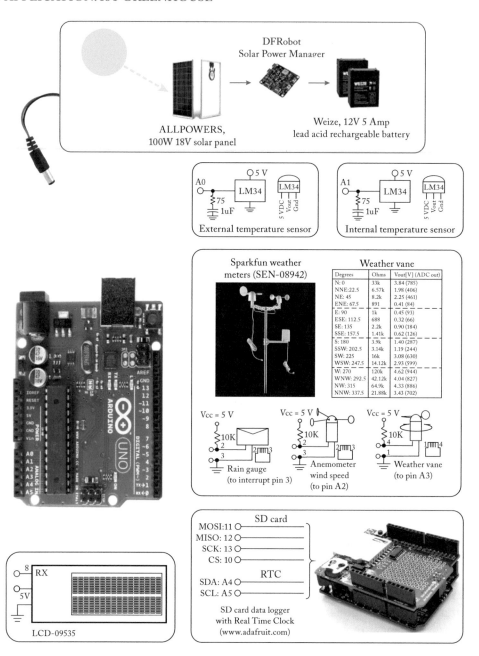

Figure 4.6: Circuit diagram for weather station. Illustrations used with permission of Sparkfun Electronics [www.sparkfun.com] and Adafruit [www.adafruit.com]. UNO R3 illustration used with permission of the Arduino Team (CC BY-NC-SA) www.arduino.cc.

5 VDC at 5 amps and 12 VDC at 8 amps. This is more than required for the weather station; however, we expand upon this project in future volumes of this text [www.DFRobot.com].

4.6.3   BOTTOM-UP IMPLEMENTATION

A sound implementation approach is to build up a complex system a subsystem at a time. We use this approach to assemble the weather station. We start with the LCD and then develop small test programs to assemble and test each weather station subsystem.

Liquid Crystal Display.   We use Sparkfun LCD-09395 to display temperature, wind speed, wind direction, and accumulated rainfall. The following sketch may be used to display these weather parameters.

```
//*********************************************************************
//Example uses the Arduino Software Serial Library
// - provides software-based serial port
//Sparkfun LCD-09395: 2 line, 16 character LCD,
// - toggles between weather readings every 2 seconds
//*********************************************************************

#include <SoftwareSerial.h>

//Arduino pins for Serial connection:
//   RX_pin: microcontroller receive pin
//   TX_pin: microcontroller transmit pin

SoftwareSerial LCD(8, 9);

void setup()
{
LCD.begin(9600);                    //Baud rate: 9600 Baud
delay(500);                         //Delay for display
}

void loop()
{
//Cursor to line one, character one
LCD.write(254);
LCD.write(128);
```

```
//clear display
LCD.write("                 ");
LCD.write("                 ");

//Cursor to line one, character one
LCD.write(254);
LCD.write(128);
LCD.write("Temp:");

//Cursor to line two, character one
LCD.write(254);
LCD.write(192);
LCD.write("Speed:");

delay(2000);                        //delay 2s

//Cursor to line one, character one
LCD.write(254);
LCD.write(128);

//clear display
LCD.write("                 ");
LCD.write("                 ");

//Cursor to line one, character one
LCD.write(254);
LCD.write(128);
LCD.write("Dir:");

//Cursor to line two, character one
LCD.write(254);
LCD.write(192);
LCD.write("Rain:");

delay(2000);                        //delay 2s

}
//***************************************************************
```

Temperature Sensors. The weather station uses two LM34 Precision Fahrenheit Temperature Sensors manufactured by Texas Instruments. The LM34 provides 10 mV of output per degree Fahrenheit. The output (center) pin of the LM34 is provided to analog input pin A0 on the Arduino UNO R3 for external temperature readings. The second LM34 is connected to pin A1 for internal temperatures. Provided below is test code to measure the output from the LM34, convert the LM34 output to temperature, and display the result on the LCD. It can be readily adapted for use on channel A1.

```
//**********************************************************************
//Example uses the Arduino Software Serial Library
// - provides software-based serial port
//**********************************************************************

#include <SoftwareSerial.h>

//Specify Arduino pins for Serial connection:
//   SoftwareSerial LCD(RX_pin, TX_pin);
SoftwareSerial LCD(8, 9);

//analog input pins
#define LM34_sensor      A0          //LM34 temp sensor at A0


int analog_temp;
int int_temp;
int troubleshoot = 1;               //1: serial monitor prints
char tempstring[10];                //create string array for LCD

void setup()
{
if (troubleshoot == 1) Serial.begin(9600);
LCD.begin(9600);                    //Baud rate: 9600 Baud
delay(500);                         //Delay for display
analogReference(DEFAULT);           //Reference voltage for ADC
}

void loop()
{                                   //Read temp from LM34 at A0
analog_temp = analogRead(LM34_sensor);
```

```
if (troubleshoot == 1) Serial.println (analog_temp);
                                //LM34 10mV/degree
int_temp = (int)(((analog_temp/1023.0) * 5.0)/.010);
if (troubleshoot == 1) Serial.println (int_temp);
sprintf(tempstring, "%4d", int_temp); //create string array for LCD

//Cursor to line one, character one
LCD.write(254);
LCD.write(128);

//clear display
LCD.write("                ");
LCD.write("                ");

//Cursor to line one, character one
LCD.write(254);
LCD.write(128);

LCD.write("Temp [F]:");

//Cursor to line one, character ten
LCD.write(254);
LCD.write(137);
LCD.write(tempstring);    //write temp string

delay(2000);
}


//*****************************************************************
```

Wind Direction Display. Provided below is an Arduino sketch to measure the output voltage from the weather vane, convert the voltage to a corresponding wind direction, and display the result to the LCD.

```
//*****************************************************************
//Example uses the Arduino Software Serial Library
// - provides software-based serial port
//*****************************************************************
```

```
#include <SoftwareSerial.h>
//Specify Arduino pins for Serial connection:
// SoftwareSerial LCD(RX_pin, TX_pin);

SoftwareSerial LCD(8, 9);
//analog input pins
#define wind_vane_sensor A3         //wind vane sensor

int analog_dir;
int int_dir;
float wind_from_dir;
int troubleshoot = 1;               //1: serial monitor prints

void setup()
{
if (troubleshoot == 1) Serial.begin(9600);
LCD.begin(9600);                    //Baud rate: 9600 Baud
delay(500);                         //Delay for display
analogReference(DEFAULT);           //Reference voltage for ADC
}

void loop()
{
                                    //Read dir from vane at A1
analog_dir = analogRead(wind_vane_sensor);
if (troubleshoot == 1) Serial.println (analog_dir);
                                    //Cursor to line two, char one
LCD.write(254);
LCD.write(192);
//clear display
LCD.write("                ");
LCD.write("                ");
                                    //Cursor to line two, char one
LCD.write(254);
LCD.write(192);
LCD.write("Dir:");
                                    //Cursor to line two, char ten
LCD.write(254);
```

```
LCD.write(202);

//Determine wind direction - vane 0 degree aligned with North
//North - ADC output 785
if((analog_dir >= 744) && (analog_dir <= 806))
  {
  char dirstring[4] = "N  ";
  LCD.write(dirstring);
  wind_from_dir = 0.0;
  if (troubleshoot == 1) Serial.println (wind_from_dir);
  }

//NNE - ADC output 406
else if((analog_dir >= 347) && (analog_dir <= 433))
  {
  char dirstring[4] = "NNE";
  LCD.write(dirstring);
  wind_from_dir = 22.5;
  if (troubleshoot == 1) Serial.println (wind_from_dir);
  }

//NE - ADC output 461
else if((analog_dir >= 434) && (analog_dir <= 530))
  {
  char dirstring[4] = "NE ";
  LCD.write(dirstring);
  wind_from_dir = 45.0;
  if (troubleshoot == 1) Serial.println (wind_from_dir);
  }

//ENE - ADC output 84
else if((analog_dir >= 76) && (analog_dir <= 88))
  {
  char dirstring[4] = "ENE";
  LCD.write(dirstring);
  wind_from_dir = 67.5;
  if (troubleshoot == 1) Serial.println (wind_from_dir);
  }
```

```
//E - ADC output 93
else if((analog_dir >= 89) && (analog_dir <= 109))
  {
  char dirstring[4] = "E  ";
  LCD.write(dirstring);
  wind_from_dir = 90.0;
  if (troubleshoot == 1) Serial.println (wind_from_dir);
  }

//ESE - ADC output 66
else if((analog_dir >= 62) && (analog_dir <= 68))
  {
  char dirstring[4] = "ESE";
  LCD.write(dirstring);
  wind_from_dir = 112.5;
  if (troubleshoot == 1) Serial.println (wind_from_dir);
  }

//SE - ADC output 184
else if((analog_dir >= 181) && (analog_dir <= 187))
  {
  char dirstring[4] = "SE";
  LCD.write(dirstring);
  wind_from_dir = 135.0;
  if (troubleshoot == 1) Serial.println (wind_from_dir);
  }

//SSE - ADC output 126
else if((analog_dir >= 110) && (analog_dir <= 155))
  {
  char dirstring[4] = "SSE";
  LCD.write(dirstring);
  wind_from_dir = 157.5;
  if (troubleshoot == 1) Serial.println (wind_from_dir);
  }

//S - ADC output 287
else if((analog_dir >= 266) && (analog_dir <= 346))
```

```
    {
    char dirstring[4] = "S  ";
    LCD.write(dirstring);
    wind_from_dir = 180.0;
    if (troubleshoot == 1) Serial.println (wind_from_dir);
    }

//SSW - ADC output 244
else if((analog_dir >= 214) && (analog_dir <= 265))
    {
    char dirstring[4] = "SSW";
    LCD.write(dirstring);
    wind_from_dir = 202.5;
    if (troubleshoot == 1) Serial.println (wind_from_dir);
    }

//SW - ADC output 630
else if((analog_dir >= 615) && (analog_dir <= 665))
    {
    char dirstring[4] = "SW ";
    LCD.write(dirstring);
    wind_from_dir = 225;
    if (troubleshoot == 1) Serial.println (wind_from_dir);
    }

//WSW - ADC output 599
else if((analog_dir >= 531) && (analog_dir <= 614))
    {
    char dirstring[4] = "WSW";
    LCD.write(dirstring);
    wind_from_dir = 247.5;
    if (troubleshoot == 1) Serial.println (wind_from_dir);
    }

//W - ADC output 944
else if((analog_dir >= 916) && (analog_dir <= 1023))
    {
    char dirstring[4] = "W  ";
```

```
  LCD.write(dirstring);
  wind_from_dir = 270.0;
  if (troubleshoot == 1) Serial.println (wind_from_dir);
  }

//WNW - ADC output 827
else if((analog_dir >= 807) && (analog_dir <= 856))
  {
  char dirstring[4] = "WNW";
  LCD.write(dirstring);
  wind_from_dir = 292.5;
  if (troubleshoot == 1) Serial.println (wind_from_dir);
  }

//NW - ADC output 886
else if((analog_dir >= 857) && (analog_dir <= 915))
  {
  char dirstring[4] = "NW ";
  LCD.write(dirstring);
  wind_from_dir = 315.0;
  if (troubleshoot == 1) Serial.println (wind_from_dir);
  }

//NNW - ADC output 702
else if((analog_dir >= 667) && (analog_dir <= 743))
  {
  char dirstring[4] = "NNW";
  LCD.write(dirstring);
  wind_from_dir = 337.5;
  if (troubleshoot == 1) Serial.println (wind_from_dir);
  }

else
  {
  char dirstring[4] = "<->";
  LCD.write(dirstring);
  }
```

```
//LCD.write(dirstring);
delay(500);
}


//********************************************************************
```

Rain Gauge. The rain gauge portion of the Sparkfun Weather Meters (SEN-08942) provides a switch closure for each tip of the rain bucket. When the bucket is full, it tips, tripping a magnetic read switch. The switch closure is routed via an RJ11 connector to Arduino interrupt pin 3. Each tip represents 0.011 inch (0.2794 mm) of rainfall. In the Arduino sketch provided below, the interrupt increments a switch closure counter and converts the result to accumulated rainfall.

```
//********************************************************************
//Program measures the accumulated rainfall since the last reset.
//The rain gauge switch is in series with a 10K resistor pulled up
//to 5 VDC.  The switch is provided to INT1 (pin 3) of the UNO R3.
//********************************************************************

unsigned long rain_sw_closures = 0;
float rainfall;

void setup()
{
Serial.begin(9600);
pinMode(3, INPUT);
attachInterrupt(1, int1_ISR, FALLING);
}


void loop()
{

//wait for interrupts

}


//********************************************************************
//int_ISR: interrupt service routine for INT1
```

```
//******************************************************************

void int1_ISR(void)
{
rain_sw_closures++;                         //increment rainfall count
                                            //switch closures to inches
rainfall = ((float)(rain_sw_closures) * 0.011);
Serial.print(rainfall);
Serial.println("   inches");
Serial.println();
}


//******************************************************************
```

Anemometer-wind Speed.   The anemometer portion of the Sparkfun Weather Meters (SEN-08942) is equipped with a three cups to catch the wind. With each rotation of the anemometer, a reed switch is tripped. Switch closures one second apart equates to 1.492 miles per hour (2.4 km/h) of wind speed. The switch closure is routed via an RJ11 connector to Arduino interrupt pin 2. In the Arduino sketch provided below, the time between two switch closures is measured and converted to wind speed in MPH.

```
//******************************************************************
//Program measures the elapsed time in ms between two switch
//closures of the anemometer.  The anemometer switch is in series
//with a 10K resistor pulled up to 5 VDC.  The switch is provided
//to INT0 (pin 2) of the UNO R3.
//******************************************************************

unsigned long first, second, elapsed_time;       //milliseconds
unsigned int first_time_hack = 1, wind_speed;

void setup()
{
Serial.begin(9600);
pinMode(2, INPUT);
attachInterrupt(0, int0_ISR, FALLING);
}
```

```
void loop()
{

//wait for interrupts

}


//*******************************************************************
//int0_ISR: interrupt service routine for INT0
//*******************************************************************

void int0_ISR(void)
{
if(first_time_hack ==1)
  {
  first = millis();                                     //milliseconds
  first_time_hack = 0;
  }
else
  {
  second = millis();                                    //milliseconds
  first_time_hack = 1;
  elapsed_time = second-first;                          //milliseconds
                                                        //ms to MPH
  wind_speed = (unsigned int)((1000.0/(float)(elapsed_time)) * 1.492);
  Serial.print(wind_speed);
  Serial.println(" MPH");
  Serial.println();
  }
}


//*******************************************************************
```

In a multiple interrupt system or a program with multiple time delays, the interrupts from the anemometer may be missed. This will result in incorrect readings. As an alternative a non-interrupt version may be used. Provided below is a sketch that uses analog input A2 to measure wind speed.

```
//*********************************************************************
//wind_speed
//*********************************************************************

#define windspeedpin    A2

unsigned long first_edge, second_edge;
unsigned long elapsed_time, start_time, wait_time, time_hack;
unsigned int  wind_speed, troubleshoot = 1;


void setup()
{
Serial.begin(9600);
pinMode(windspeedpin, INPUT);
}

void loop()
{
start_time = millis();
wait_time = 0;
if(troubleshoot)
  {
  Serial.print("Wait Time initial:");
  Serial.print(wait_time);
  Serial.println();
  }                                        //process pin while low
while((analogRead(windspeedpin) > 512) && (wait_time < 1500))
  {
  time_hack = millis();
  wait_time = time_hack - start_time;
  if(troubleshoot)
    {
    Serial.print("Wait Time low1.........................:");
    Serial.print(wait_time);
    Serial.println();
    }
  }
```

```
start_time = millis();
wait_time = 0;                          //process pin while high
while((analogRead(windspeedpin) >= 512) && (wait_time < 1500))
  {
  time_hack = millis();
  wait_time = time_hack - start_time;
  if(troubleshoot)
    {
    Serial.print("Wait Time high:");
    Serial.print(wait_time);
    Serial.println();
    }
  }
first_edge = millis();                  //capture first edge

if(troubleshoot)
    {
    Serial.print("First edge:");
    Serial.print(first_edge);
    Serial.println();
    }


start_time = millis();
wait_time = 0;                          //process pin while low
while((analogRead(windspeedpin) < 512) && (wait_time < 1500))
  {
  time_hack = millis();
  wait_time = time_hack - start_time;
  if(troubleshoot)
    {
    Serial.print("Wait Time low2...................:");
    Serial.print(wait_time);
    Serial.println();
    }
  }
second_edge = millis();                 //capture second edge
```

```
if(troubleshoot)
    {
    Serial.print("Second edge:");
    Serial.print(second_edge);
    Serial.println();
    }

elapsed_time = second_edge - first_edge;//elapsed time

if(troubleshoot)
    {
    Serial.print("Time between edges:");
    Serial.print(elapsed_time);
    Serial.println();
    }
                                        //convert to wind speed
wind_speed = (unsigned int)((1000.0/(float)(elapsed_time)) * 1.492);

Serial.print("MPH:");
Serial.print(wind_speed);
Serial.println();
Serial.println();

delay(1000);
}


//*****************************************************************
```

SD Card with Real Time Clock.   We use the Adafruit data logger shield (Adafruit #1141). It is equipped with a RTC and battery backup to timestamp collected data. Adafruit provides step-by-step instructions to connect header pins and program the SD Card shield in "Adafruit Data Logger Shield [www.adafruit.com]." We adapted the code to log data in the final weather station system code provided later in the chapter.

Solar Power System.   The solar power system for the remote weather station is shown in Figure 4.7. The system is managed by the DFRobot DFR0590 Solar Power Management Module. This module was designed for use with an 18 V, 100 W solar panel and provides trickle charging for a 12 V lead-acid battery. The solar panel chosen for the system is an All Powers 18 V, 100 W solar panel (#AP-SP-016-SIL). It has a working voltage of 18 V and working current of 5.8 A.

The solar panel is connected to the DFR0590 Solar Power Management Module via 10 AWG insulated power cables equipped with MC4 connectors. The battery chosen for the system is the Weize FP1250 12V, 5AH sealed lead-acid rechargeable battery. The power cable from the solar panel to the power management module is fused at 7.5 A. The power cable between the power management module and the battery is fused at 5 A. The solar power system is more than adequate to meet the needs of the remote weather station. We add to the project in future volumes of the textbook series.[3]

4.6.4   UML ACTIVITY DIAGRAM

The UML activity diagram for the main program is provided in Figure 4.8. After initializing the subsystems, the program enters a continuous loop where temperature and wind direction is sensed and displayed on the LCD and the LED display. Interrupts are used to capture data on wind speed and rainfall. The sensed values are then transmitted via the SPI to the MMC/SD card. The system then enters a delay to set how often the temperature and wind direction parameters are updated. We have you construct the individual UML activity diagrams for each function as an end of chapter exercise.

4.6.5   MICROCONTROLLER CODE

The final code may be assembled from all of the individual code sketches provided in this section. It is highly recommended that the code be brought up a portion at a time.

---

[3]Morgan & Claypool is a global publisher. Due to the wide range of applicable codes and standards and permitting requirements, please check with local requirements concerning solar panel installation.

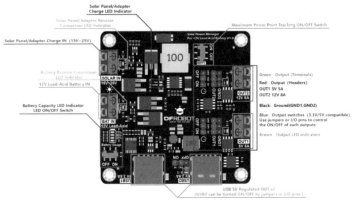

(a) DF Robot DFR0580 Solar Power Management Module.
Illustration courtesy of DF Robot [www.dfrobot.com].

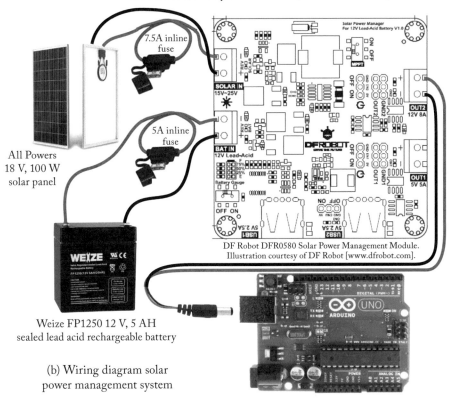

(b) Wiring diagram solar
power management system

Figure 4.7: Solar power system for the remote weather station. Illustrations courtesy of DFRobot [www.DFRobot.com], All Powers [www.allpowers.com], and Weize [www.weize.com].

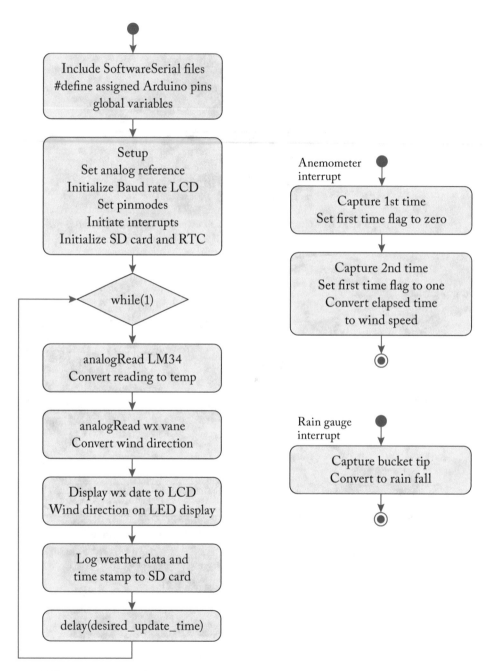

Figure 4.8: Weather station UML activity diagram.

```
//**********************************************************************
//Senses external and internal greenhouse parameters.
//Logs time stamp and parameters to SD card
//
//https://github.com/adafruit/Light-and-Temp-logger/
//Originally developed by Lady Ada
//Adapted for six sensors
//**********************************************************************
                                       //include libraries
#include <SPI.h>                       //SPI library
#include <SD.h>                        //SD card library
#include <Wire.h>                      //Wire library
#include "RTClib.h"                    //Real Time Clock library
#include <SoftwareSerial.h>            //Software Serial Library


                                       //#define statements
#define LOG_INTERVAL   1000            //ms between capture events
                                       //ms between logging events
#define SYNC_INTERVAL 1000             //e.g., set 10*LOG_INTERVAL
                                       //to write
#define ECHO_TO_SERIAL   0             //echo data to serial port
#define WAIT_TO_START    0             //wait for serial input
                                       // in setup()
#define ext_tempPin      A0            //analog 0: external temp
#define int_tempPin      A1            //analog 1: internal temp
#define windspeedpin     A2            //analog 2: wind speed
#define wind_vane_sensor A3            //analog 3: wind vane
                                       //global variables
uint32_t syncTime = 0;                 //time of last sync()
const int chipSelect = 10;             //D10 for SD card CS line
char tempstring[10];                   //string array for LCD
unsigned int wind_from_dir_degrees;    //wind from dir specified
                                       //in degrees
File logfile;                          //logging file

unsigned int  first_edge, second_edge, elapsed_time;
unsigned int  start_time, wait_time, time_hack;
unsigned int  wind_speed;
```

```
unsigned int rain_sw_closures = 0;
float rainfall;
                                        //initialize LCD, RTC
SoftwareSerial LCD(8, 9);               //SoftwareSerial
                                        // (RX_pin, TX_pin)
RTC_PCF8523 RTC;                        //Real Time Clock object

void setup(void)
{
pinMode(3, INPUT);                      //attach rainfall interrupt
attachInterrupt(1, int1_ISR, FALLING);

#if ECHO_TO_SERIAL
  Serial.begin(9600);                   //configure serial monitor
  Serial.println();
#endif

LCD.begin(9600);                        //configure LCD Baud: 9600
delay(500);                             //delay for display

#if WAIT_TO_START
  Serial.println("Type any character to start");
  while (!Serial.available());
#endif //WAIT_TO_START
                                        //initializing SD card
#if ECHO_TO_SERIAL
  Serial.print("Initializing SD card...");
#endif

pinMode(10, OUTPUT);                    //configure CS pin

if(!SD.begin(chipSelect))               //card present?
  {
  error("Card failed, or not present");
  }

#if ECHO_TO_START
  Serial.println("card initialized.");
```

```
#endif

char filename[] = "LOGGER00.CSV";            //create a new file
for(uint8_t i = 0; i < 100; i++)
  {
  filename[6] = i/10 + '0';
  filename[7] = i%10 + '0';
  if(! SD.exists(filename))
    {                                        //open new
    logfile = SD.open(filename, FILE_WRITE);
    break;  // leave the loop!
    }
  }

if(! logfile)
  {
  error("could not create file");
  }

#if ECHO_TO_SERIAL
  Serial.print("Logging to: ");
  Serial.println(filename);
#endif


                                             //Real Time Clock
Wire.begin();                                //connect to RTC
if(!RTC.begin())
  {
  logfile.println("RTC failed");
  #if ECHO_TO_SERIAL
    Serial.println("RTC failed");
  #endif  //ECHO_TO_SERIAL
  }
                                             //to logfile
logfile.println("millis,stamp,datetime,exttemp,inttemp,windspeed,"
                "winddir, rainfall");
#if ECHO_TO_SERIAL
  Serial.println("millis,stamp,datetime,exttemp,inttemp,windspeed,"
```

```
                    "winddir, rainfall");
#endif //ECHO_TO_SERIAL
}

void loop(void)
{
DateTime now;                                  //time variable
                                               //delay between readings
delay((LOG_INTERVAL -1) - (millis() % LOG_INTERVAL));
uint32_t m = millis();                         //log milliseconds since
                                               // starting

logfile.print(m);                              //ms since start
logfile.print(", ");
#if ECHO_TO_SERIAL
  Serial.print(m);                             //ms since start
  Serial.print(", ");
#endif

now = RTC.now();                               //fetch the time
                                               //log time
logfile.print(now.unixtime());                 //seconds since 1/1/1970
  logfile.print(", ");                logfile.print('"');
  logfile.print(now.year(), DEC);     logfile.print("/");
  logfile.print(now.month(), DEC);    logfile.print("/");
  logfile.print(now.day(), DEC);      logfile.print(" ");
  logfile.print(now.hour(), DEC);     logfile.print(":");
  logfile.print(now.minute(), DEC);   logfile.print(":");
  logfile.print(now.second(), DEC);   logfile.print('"');
#if ECHO_TO_SERIAL
  Serial.print(now.unixtime());                //seconds since 1/1/1970
  Serial.print(", ");                 Serial.print('"');
  Serial.print(now.year(), DEC);      Serial.print("/");
  Serial.print(now.month(), DEC);     Serial.print("/");
  Serial.print(now.day(), DEC);       Serial.print(" ");
  Serial.print(now.hour(), DEC);      Serial.print(":");
  Serial.print(now.minute(), DEC);    Serial.print(":");
  Serial.print(now.second(), DEC);    Serial.print('"');
#endif //ECHO_TO_SERIAL
```

```
//collect data for all sensors
analogRead(ext_tempPin);                          //A0: external temp
delay(10);
int ext_temp_Reading = analogRead(ext_tempPin);

analogRead(int_tempPin);                          //A1: internal temp
delay(10);
int int_temp_Reading = analogRead(int_tempPin);

//convert reading to voltage, based off the reference voltage
unsigned int ext_temperatureF =
  (unsigned int)((float)(ext_temp_Reading) * 5.0/1024.0 * 100);
unsigned int int_temperatureF =
  (unsigned int) ((float)(int_temp_Reading) * 5.0/1024.0  * 100);

//display data sequentially to LCD
clear_LCD();
//A0: outdoor temp, A1: indoor temp
char buff0[5] = "ExtF";                           //ext temp to LCD line1
LCD_print_line1((int)(ext_temperatureF), buff0);

char buff1[5] = "IntF";                           //int temp to LCD line2
LCD_print_line2((int)(int_temperatureF), buff1);

delay(200);                                       //persistence delay
clear_LCD();                                       //clear LCD

//<<<<<<<
start_time = millis();
wait_time = 0;

#if ECHO_TO_SERIAL
  Serial.print("Wait Time initial:");
  Serial.print(wait_time);
  Serial.println();
#endif //ECHO_TO_SERIAL

  //process pin while low
```

```
while((analogRead(windspeedpin) > 512) && (wait_time < 1500))
  {
  time_hack = millis();
  wait_time = time_hack - start_time;
  #if ECHO_TO_SERIAL
    Serial.print("Wait Time low1........................:");
    Serial.print(wait_time);
    Serial.println();
  #endif //ECHO_TO_SERIAL
  }

start_time = millis();
wait_time = 0;                                  //process while high
while((analogRead(windspeedpin) >= 512) && (wait_time < 1500))
  {
  time_hack = millis();
  wait_time = time_hack - start_time;
  #if ECHO_TO_SERIAL
    Serial.print("Wait Time high:");
    Serial.print(wait_time);
    Serial.println();
  #endif //ECHO_TO_SERIAL
  }
first_edge = millis();                          //capture first edge

#if ECHO_TO_SERIAL
  Serial.print("First edge:");
  Serial.print(first_edge);
  Serial.println();
#endif //ECHO_TO_SERIAL

start_time = millis();
wait_time = 0;                                  //process while low
while((analogRead(windspeedpin) < 512) && (wait_time < 1500))
  {
  time_hack = millis();
  wait_time = time_hack - start_time;
  #if ECHO_TO_SERIAL
```

```
     Serial.print("Wait Time low2...................:");
     Serial.print(wait_time);
     Serial.println();
  #endif //ECHO_TO_SERIAL
  }
second_edge = millis();                              //capture 2nd edge

#if ECHO_TO_SERIAL
   Serial.print("Second edge:");
   Serial.print(second_edge);
   Serial.println();
#endif //ECHO_TO_SERIAL

elapsed_time = second_edge - first_edge;//elapsed time

#if ECHO_TO_SERIAL
   Serial.print("Time between edges:");
   Serial.print(elapsed_time);
   Serial.println();
#endif //ECHO_TO_SERIAL
                                                     //convert wind sp
wind_speed = (unsigned int)((1000.0/(float)(elapsed_time)) * 1.492);

#if ECHO_TO_SERIAL
   Serial.print("MPH:");
   Serial.print(wind_speed);
   Serial.println();
   Serial.println();
 #endif //ECHO_TO_SERIAL

//A2: wind speed
char buff2[11] = "SpeedMPH";                         //speed to line 1
LCD_print_line1((int)(wind_speed), buff2);

//<<<<<<<

//delay(200);                                        //persistence delay
wind_from_dir_degrees = get_wind_direction();
```

```
delay(200);                                    //persistence delay
clear_LCD();                                   //clear both lines

char buff3[11] = "Rainfall";                   //rainfall line 1
LCD_print_line1((int)(rainfall * 1000.0), buff3);

delay(200);                                    //persistence delay
clear_LCD();                                   //clear both lines

#if ECHO_TO_SERIAL
   Serial.println (wind_from_dir_degrees);
#endif //ECHO_TO_SERIAL

//log data to SD card
logfile.print(", ");                           //log data
logfile.print(ext_temperatureF);
logfile.print(", ");
logfile.print(int_temperatureF);
logfile.print(", ");
logfile.print(wind_speed);
logfile.print(", ");
logfile.print(wind_from_dir_degrees);
logfile.print(", ");
logfile.print(rainfall);

#if ECHO_TO_SERIAL                             //to serial monitor
  Serial.print(", ");
  Serial.print(ext_temperatureF);
  Serial.print(", ");
  Serial.print(int_temperatureF);
  Serial.print(", ");
  Serial.print(wind_speed);
  Serial.print(", ");
  Serial.print(wind_from_dir_degrees);
  Serial.print(", ");
  Serial.print(rainfall);
#endif //ECHO_TO_SERIAL
```

```
logfile.println();
#if ECHO_TO_SERIAL
  Serial.println();
#endif // ECHO_TO_SERIAL

if ((millis() - syncTime) < SYNC_INTERVAL) return;    //write to disk!
syncTime = millis();
logfile.flush();                                      //sync data to SD
                                                      //and update FAT

}

//*********************************************************************

void error(char *str)
{
Serial.print("error: ");
Serial.println(str);

while(1);
}

//*********************************************************************

void LCD_print_line1(int result1, char *ptr)
{
//print result to LCD via SoftwareSerial
sprintf(tempstring, "%4d", (result1)); //create string array for LCD
//Cursor to line one, character one
LCD.write(254);
LCD.write(128);
//clear display
LCD.write("                 ");
//LCD.write(" ");
//Cursor to line one, character one
LCD.write(254);
LCD.write(128);
LCD.write(ptr);
//Cursor to line one, character thirteen
```

```
LCD.write(254);
LCD.write(140);
LCD.write(tempstring); //write temp string
delay(200);
}


//********************************************************************


void LCD_print_line2(int result1, char *ptr)
{
//print result to LCD via SoftwareSerial
sprintf(tempstring, "%4d", (result1)); //create string array for LCD
//Cursor to line two, character one
LCD.write(254);
LCD.write(192);
//clear display
LCD.write("                ");
//LCD.write(" ");
//Cursor to line two, character one
LCD.write(254);
LCD.write(192);
LCD.write(ptr);
//Cursor to line two, character thirteen
LCD.write(254);
LCD.write(204);
LCD.write(tempstring); //write temp string
delay(200);
}


//********************************************************************


void clear_LCD(void)
{
//Cursor to line one, character one
LCD.write(254);
LCD.write(128);
//clear display
LCD.write("                ");
```

```
//Cursor to line two, character one
LCD.write(254);
LCD.write(192);
//clear display
LCD.write("                  ");

delay(20);                                              //for LCD
}


//*********************************************************************

unsigned int get_wind_direction(void)
{
int analog_dir;
unsigned int wind_from_dir;

analog_dir = analogRead(wind_vane_sensor);              //Read dir vane A1
                                                        //Cursor to line two
                                                        //character one
LCD.write(254);
LCD.write(192);
LCD.write("Dir:");

                                                        //Cursor to line two
                                                        //character ten
LCD.write(254);
LCD.write(205);

//Determine wind direction - vane 0 degree aligned with North
//North - ADC output 785
if((analog_dir >= 744) && (analog_dir <= 806))
  {
  char dirstring[4] = "-N-";
  LCD.write(dirstring);
  wind_from_dir = 0;
  }

//NNE - ADC output 406
else if((analog_dir >= 347) && (analog_dir <= 433))
```

```
  {
  char dirstring[4] = "NNE";
  LCD.write(dirstring);
  wind_from_dir = 22;
  }


//NE - ADC output 461
else if((analog_dir >= 434) && (analog_dir <= 530))
  {
  char dirstring[4] = "N-E";
  LCD.write(dirstring);
  wind_from_dir = 45;
  }


//ENE - ADC output 84
else if((analog_dir >= 76) && (analog_dir <= 88))
  {
  char dirstring[4] = "ENE";
  LCD.write(dirstring);
  wind_from_dir = 67;
  }


//E - ADC output 93
else if((analog_dir >= 89) && (analog_dir <= 109))
  {
  char dirstring[4] = "-E-";
  LCD.write(dirstring);
  wind_from_dir = 90;
  }


//ESE - ADC output 66
else if((analog_dir >= 62) && (analog_dir <= 68))
  {
  char dirstring[4] = "ESE";
  LCD.write(dirstring);
  wind_from_dir = 112;
  }
```

```
//SE - ADC output 184
else if((analog_dir >= 181) && (analog_dir <= 187))
  {
  char dirstring[4] = "S-E";
  LCD.write(dirstring);
  wind_from_dir = 135;
  }

//SSE - ADC output 126
else if((analog_dir >= 110) && (analog_dir <= 155))
  {
  char dirstring[4] = "SSE";
  LCD.write(dirstring);
  wind_from_dir = 157;
  }

//S - ADC output 287
else if((analog_dir >= 266) && (analog_dir <= 346))
  {
  char dirstring[4] = "-S-";
  LCD.write(dirstring);
  wind_from_dir = 180;
  }

//SSW - ADC output 244
else if((analog_dir >= 214) && (analog_dir <= 265))
  {
  char dirstring[4] = "SSW";
  LCD.write(dirstring);
  wind_from_dir = 202;
  }

//SW - ADC output 630
else if((analog_dir >= 615) && (analog_dir <= 665))
  {
  char dirstring[4] = "S-W";
  LCD.write(dirstring);
  wind_from_dir = 225;
```

```
      }

//WSW - ADC output 599
else if((analog_dir >= 531) && (analog_dir <= 614))
  {
  char dirstring[4] = "WSW";
  LCD.write(dirstring);
  wind_from_dir = 247;
  }

//W - ADC output 944
else if((analog_dir >= 916) && (analog_dir <= 1023))
  {
  char dirstring[4] = "-W-";
  LCD.write(dirstring);
  wind_from_dir = 270;
  }

//WNW - ADC output 827
else if((analog_dir >= 807) && (analog_dir <= 856))
  {
  char dirstring[4] = "WNW";
  LCD.write(dirstring);
  wind_from_dir = 292;
  }

//NW - ADC output 886
else if((analog_dir >= 857) && (analog_dir <= 915))
  {
  char dirstring[4] = "N-W";
  LCD.write(dirstring);
  wind_from_dir = 315;
  }

//NNW - ADC output 702
else if((analog_dir >= 667) && (analog_dir <= 743))
  {
  char dirstring[4] = "NNW";
```

```
    LCD.write(dirstring);
    wind_from_dir = 337;
    }

else
    {
    char dirstring[4] = "<->";
    LCD.write(dirstring);
    }

delay(200);

return wind_from_dir;
}


//*******************************************************************
//int_ISR: interrupt service routine for INT1
//Program measures the accumulated rainfall since the last reset.
//The rain gauge switch is in series with a 10K resistor pulled up
//to 5 VDC.  The switch is provided to INT1 (pin 3) of the UNO R3.
//*******************************************************************

void int1_ISR(void)
{
rain_sw_closures++;                          //increment rainfall count
                                             //closures to inches
rainfall = ((float)(rain_sw_closures) * 0.011);
//Serial.print(rainfall);
//Serial.println("   inches");
//Serial.println();
}


//*******************************************************************
```

4.6.6   FINAL ASSEMBLY

The weather station may be assembled into a small package by using an Adafruit Wing Shield (#196). The Wing Shield couples to the Arduino UNO R3 with onboard stacking headers. The Adafruit data logger shield (#1141) is then stacked upon the Wing Shield. The Wing Shield is

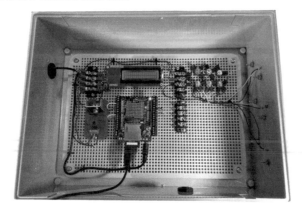

Figure 4.9: Assembled weather station. Arduino illustrations used with permission of the Arduino Team (CC BY-NC-SA) [www.arduino.cc].

equipped with screw terminals to connect the weather station peripherals to the stacked modules. The completed weather station is mounted within a QILIPSU hinged cover, stainless-steel latch, junction box with mounting plate. The final result is provided in Figure 4.9.

4.7   GREENHOUSE CONTROL

As shown in Figure 4.2, the following requirements have been set for the Greenhouse Control portion of the GIS system:

- monitor the water level in the rain barrel;

- monitor indoor greenhouse temperature;

- monitor external greenhouse temperature;

- monitor humidity level within the greenhouse;

- monitor plant soil moisture;

- activate vent fan when the internal greenhouse temperature is above a desired value;

- activate misting the system when the internal humidity level falls below the desired value; and

- monitor the voltage of the 12 VDC lead-acid battery.

To meet these requirements the instrumentation system shown in Figure 4.10 is used. Each component is discussed in turn.

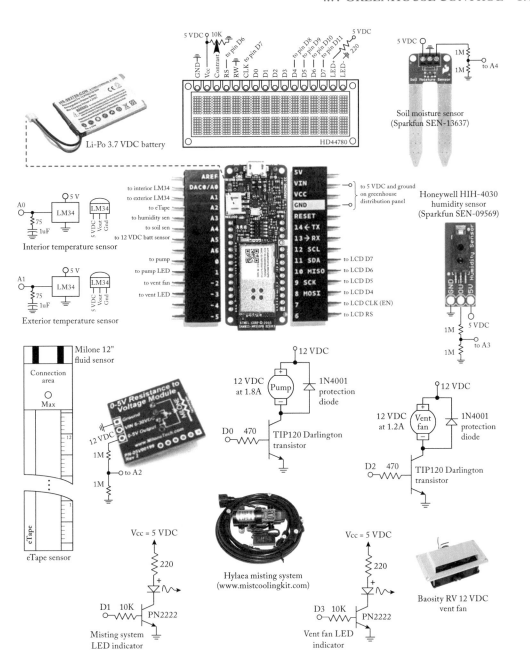

Figure 4.10: Greenhouse control system. Arduino illustrations used with permission of the Arduino Team (CC BY-NC-SA) [www.arduino.cc].

Arduino MKR1000. The MKR1000 was chosen because it is a powerful system on a chip (SOC) with a built-in WiFi module. Also, the MKR1000 IoT Bundle provides for straight forward IoT application development. The MKR1000 features a 32-bit ATSAMD21 processor operating at 48 MHz. Although the processing board is small, it is full featured with 256 KB of flash and 32 KB of RAM. The processor hosts SPI, I2C, and UART serial communication systems. The processor is also equipped with pulse width modulation (PWM) features and multiple channels of analog-to-digital conversion channels. The pinout for the MKR1000 is provided in Figure 4.11. There are several methods of providing battery power to the MKR1000 including an input connector for a 3.7 VDC rechargeable LiPo (lithium-ion polymer) battery. In this application, we provide the MKR1000 power via the VIN input.

The MKR1000 is actually a system on a chip. It uses the Microchip [www.microchip] ATSAM25 system on a chip with three onboard systems:

- SAMD 21 Cortex–M0+ 32-bit low power ARM processor,

- WINC1500 low power 2.4 GHz WiFi module and antenna, and

- ECC 508 crypto authentication unit.

Note: The MKR1000 is a 3.3 VDC system. System inputs and outputs must not exceed this value.

Li-Po Battery. The MKR1000 receives primary power from the 5 VDC, 5 A output (OUT1) of the DF Robot DFR0580 Solar Power Management Model. The 5 VDC source is routed to the MKR1000 VIN pin via the printed circuit board. The MKR1000 has a built-in recharging system for a Lithium Polymer (Li-Po) 3.7 VDC battery. A 3.7 VDC Li-Po battery with a capacity rating of 2500 mAh and equipped with a 2-pin JST connector (Jameco #2309575) is used [www.jameco.com.]

Hitachi HD44780 LCD. The MKR1000 IoT Bundle provides a number of components to assemble several IoT projects including a Hitachi HD44780 LCD. The LCD is 16 characters by 2 lines. Shown in Figure 4.12 is in the interface diagram between the MKR1000 and the LCD. The LCD contrast is set using an external 10 kOhm potentiometer. The backlit LED is biased using a 220 Ohm resistor.

The following code snapshot provides configuration and testing of the LCD using functions within the Arduino Liquid Crystal display library. This test function was adapted from the MKR1000 PuzzleBox project.

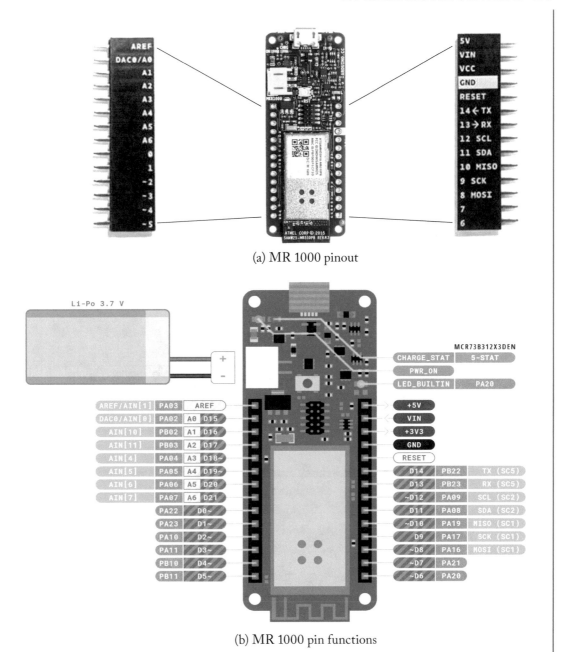

(a) MR 1000 pinout

(b) MR 1000 pin functions

Figure 4.11: **MKR1000** pinout. Illustrations used courtesy of Arduino CC-BY-SA [www. arduino.cc].

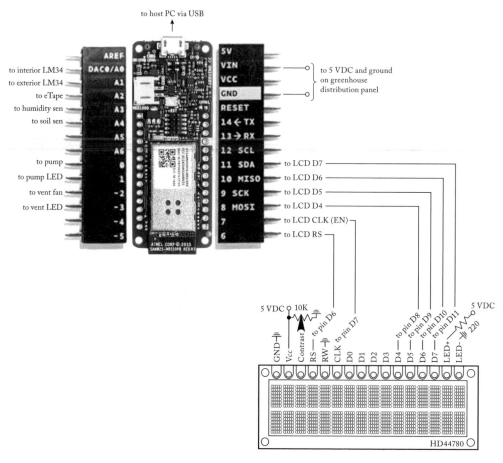

Figure 4.12: **MKR1000 to LCD interface.** Arduino illustrations used with permission of the Arduino Team (CC BY-NC-SA) [www.arduino.cc].

```
//**************************************************************
//File: LCD_44780
//Test sketch for MKR1000 to LCD HD 44780.
//LCD provided with MKR IoT Bundle
//Adapted from the MKR1000 PuzzleBox project
//Copyright CC BY-NC [Arduino.cc]
//**************************************************************

#include <LiquidCrystal.h>          //LCD support library
```

```
//associate MKR1000 pins with LCD pins
const int rs = 6, en = 7, d4 = 8, d5 = 9, d6 = 10, d7 = 11;

//initialize MKR1000 to LCD interface
LiquidCrystal lcd(rs, en, d4, d5, d6, d7);

void setup()
  {
  lcd.begin(16, 2); //LCD's number of columns and rows
  lcd.print("hello, world!"); //Print message to LCD
  }

void loop()
  {
  lcd.setCursor(0, 1);        //set cursor to column 0, line 1
  lcd.print(millis() / 1000);//print number seconds since reset
  }

//*******************************************************************
```

LM34 Interior and Exterior Greenhouse Temperature Sensor.   To monitor the interior and exterior greenhouse temperature a pair of LM34 Precision Fahrenheit Temperature sensors are used. The LM34 provides 10 mV of output per degree Fahrenheit. The interface circuitry is shown in Figure 4.13a. As configured, the LM34s report Fahrenheit temperatures down to zero degrees. A design to provide temperature sensing below zero degrees is provided as an end of chapter exercise. The LM34 measuring external temperature should be housed in a weather-protective enclosure, as shown in Figure 4.13b.

Provided below is the code snapshot to sample greenhouse temperatures via the LM34s, remap the values to an analog voltage, and to a Fahrenheit temperature.

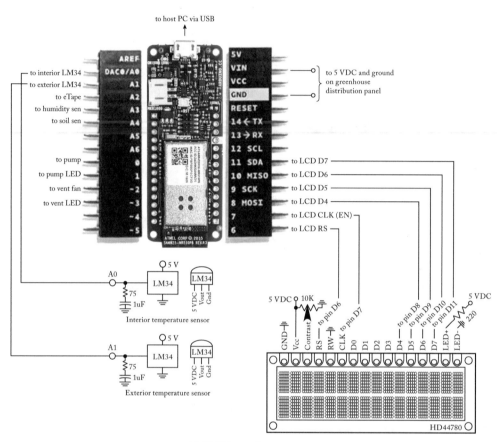

(a) LM34 interface circuitry

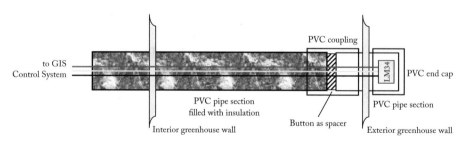

(b) Weather-protective enclosure for exterior LM34

Figure 4.13: Interior and exterior greenhouse monitor. Arduino illustrations used with permission of the Arduino Team (CC BY-NC-SA) [www.arduino.cc].

```
//*********************************************************************
//File: LCD_LM34
//Test sketch for MKR1000 to LCD HD 44780.
//LM34 connected to A0 - measures internal greenhouse temp
//LM34 connected to A1 - measures external greenhouse temp
//LCD provided with MKR IoT Bundle
//*********************************************************************

#include <LiquidCrystal.h>   //LCD support library

//associate MKR1000 pins with LCD pins
const int rs = 6, en = 7, d4 = 8, d5 = 9, d6 = 10, d7 = 11;

//initialize MKR1000 to LCD interface
LiquidCrystal lcd(rs, en, d4, d5, d6, d7);

void setup()
  {
  lcd.begin(16, 2);           //LCD's number of columns and rows
  lcd.setCursor(0, 0);        //set cursor to column 0, line 0
  lcd.print("IntTemp:");      //Print message to LCD
  lcd.setCursor(0, 1);        //set cursor to column 0, line 1
  lcd.print("ExtTemp:");      //Print message to LCD
  }

void loop()
  {
  lcd.setCursor(9, 0);        //set cursor to column 8, line 0
  int int_temp = map(analogRead(A0),0, 1023, 0, 330);
  lcd.print(int_temp);

  lcd.setCursor(9, 1);        //set cursor to column 8, line 1
  int ext_temp = map(analogRead(A1), 0, 1023, 0, 330);
  lcd.print(ext_temp);
  delay(1000);
  }

//*********************************************************************
```

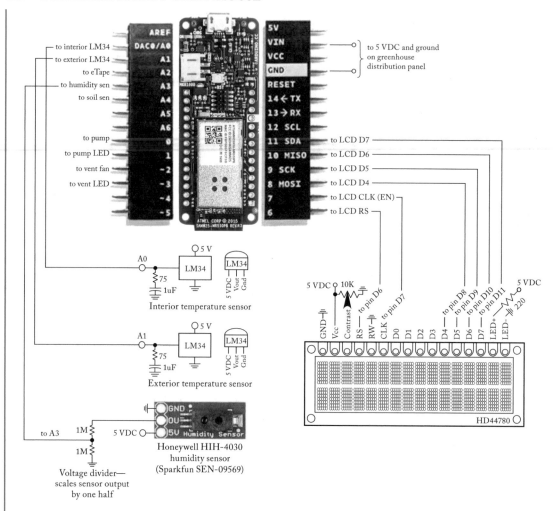

Figure 4.14: Greenhouse humidity monitoring. Arduino illustrations used with permission of the Arduino Team (CC BY-NC-SA) [www.arduino.cc]. Image of humidity sensor used courtesy of Sparkfun (CY BY 2.0) [www.sparkfun.com].

Humidity Sensor.   A Honeywell HIH-4030 sensor is used to measure greenhouse humidity. The sensor provides an output voltage that may be mapped to a corresponding relative humidity (RH) value. The RH value provides a measurement of the mount of water vapor in the air. The RH is expressed as a value from 0–100% RH. The interface circuit for the RH sensor is shown in Figure 4.14. The output from the sensor is provided to a voltage divider network of two each 1 Mohm resistors to halve the output voltage. This keeps the sensor output voltage below the 3.3 VDC input maximum of the MKR1000 analog-to-digital (ADC) system. The reading from the ADC is multiplied by a factor of two in the function below to compensate for the resistive network halving.

The sensor provides an output voltage to indicate RH. The voltage is processed and corrected for temperature using the following equations provided by the manufacturer (Honeywell [7]):

$$V_{out} = (V_{supply}) * (0.0062 * sensorRH) + 0.16).$$

The sensor RH value is corrected for temperature:

$$TrueRH = (sensorRH/(1.0546 - 0.00216T)$$

with T expressed in degrees Centigrade.

```
//***********************************************************************
//File: LCD_humidity
//Test sketch for MKR1000 to LCD HD 44780.
//LM34 connected to A0 - measures internal greenhouse temp
//LM34 connected to A1 - measures external greenhouse temp
//Honeywell HIH-4030 humidity sensor connected to A3
//   - The senor output is divided in half by resistive
//     network.  Multiply reading by two to correct.
//LCD provided with MKR IoT Bundle
//***********************************************************************

#include <LiquidCrystal.h>                //LCD support library

//associate MKR1000 pins with LCD pins
const int rs = 6, en = 7, d4 = 8, d5 = 9, d6 = 10, d7 = 11;

//initialize MKR1000 to LCD interface
LiquidCrystal lcd(rs, en, d4, d5, d6, d7);

void setup()
  {
  lcd.begin(16, 2);                 //LCD's number of columns and rows
  lcd.setCursor(0, 0);              //set cursor to column 0, line 0
  lcd.print("IntTemp:");            //Print message to LCD
  lcd.setCursor(0, 1);              //set cursor to column 0, line 1
  lcd.print("Humidity:");           //Print message to LCD
  }
```

```
void loop()
 {
 lcd.setCursor(9, 0);                  //set cursor to column 8, line 0
 int int_temp = map(analogRead(A0),0, 1023, 0, 330);
 lcd.print(int_temp);

 lcd.setCursor(9, 1);                  //set  cursor to column 9, line 1
 int humidity = analogRead(A3);   //read analog ch A3
                                  //convert reading to voltage
                                  //correct for one-half scaling
 float humidity_voltage = (float)(humidity /1023.0) * 3.3 * 2.0;
                                  //convert to RH per data sheet
 float sensor_RH_flt  = (((humidity_voltage/5.0) - 0.16) *
                         (1/0.0062));
                                  //convert temp reading to C
 float int_temp_C = (float)((int_temp - 32.0) * (5.0/9.0));
                                  //compensate for temp per data sheet
 float true_RH = sensor_RH_flt/(1.0546 - (0.00216 * int_temp_C));
 int sensor_RH = (int)(true_RH); //cast to int for display
 lcd.print(sensor_RH);
 lcd.setCursor(11, 1);                 //set cursor to column 11, line 1
 lcd.print("%");
 delay(1000);
 }

//***********************************************************************
```

Soil Moisture Sensor.   A Sparkfun soil moisture sensor (SEN-13637) is used to monitor plant soil moisture content. The sensor is powered by a 5 VDC source. To insure the sensor output does not exceed the 3.3 VDC maximum of the MKR1000, a voltage divider network is used to halve the sensor output signal. The sensor provides a low-level output voltage for dry soil and a voltage approaching the source voltage for moist soil. The interface circuit is shown in Figure 4.15. The sensor reading is multiplied by a factor of two in the sensor reading function to compensate for the resistive network halving.

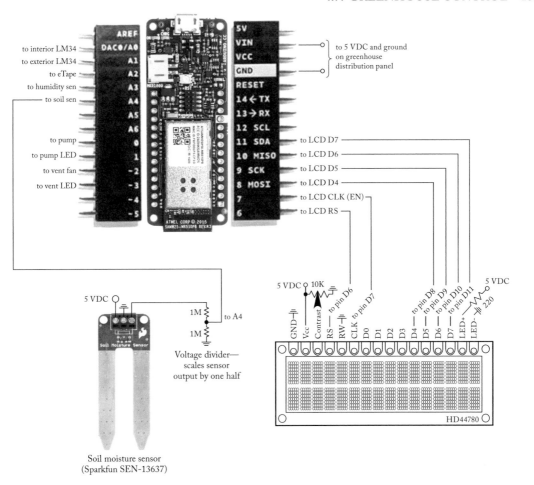

Figure 4.15: Sparkfun soil moisture sensor. Arduino illustrations used with permission of the Arduino Team (CC BY-NC-SA) [www.arduino.cc]. Image of sensor used courtesy of Sparkfun (CY BY 2.0) [www.sparkfun.com].

```
//******************************************************************
//GIS_soil_sensor:  test features of the Sparkfun soil moisture sensor
//                  SEN-13637
// - Soil sensor output is provided to A4 via a voltage divider
//   circuit.
// - The voltage divider circuit consists of a 1 Mohm and a 1 Mohm
//   resistor in series.
```

```
// - The voltage sensed at A4 is 1/2 of the soil sensor output
// - The sensed value is multiplied by 2 to compensate for the
//   resistor network scaling.
// - The input voltage to A4 is displayed on the LCD.
//*****************************************************************


#include <LiquidCrystal.h>          //LCD support library

//associate MKR1000 pins with LCD pins
const int rs = 6, en = 7, d4 = 8, d5 = 9, d6 = 10, d7 = 11;


//initialize MKR1000 to LCD interface
LiquidCrystal lcd(rs, en, d4, d5, d6, d7);

void setup()
  {
  lcd.begin(16, 2);                  //LCD's number of columns and rows
  lcd.setCursor(0, 0);               //set cursor to column 0, line 0
  lcd.print("Soil sensor:");         //Print message to LCD
  }

void loop()
  {
  lcd.setCursor(0, 1);               //set cursor to column 0, line 1
  int soil_voltage = map(analogRead(A4),0, 1023, 0, 330);
  float soil_voltage_float = (float)((soil_voltage/100.0) * 2.0);
  lcd.print(soil_voltage_float);
  delay(100);
  }


//*****************************************************************
```

Milone eTape Fluid Sensor.   Milone Technologies manufacture a line of continuous fluid level sensors. The sensor resembles a ruler and provides a near linear response. The sensor reports a change in resistance to indicate the distance from sensor top to the fluid surface. To convert the resistance change to a voltage change, the Milone 0–5 VDC Resistance to Voltage Module is used. The module shown in Figure 4.16 is powered from 12 VDC. The output from the module ranges up to 5 VDC. To insure compatibility with the MKR1000, the module output

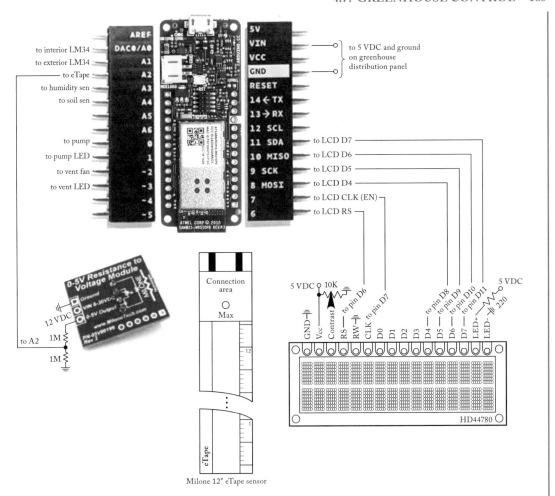

Figure 4.16: Milone Technologies eTape liquid level sensor. Arduino illustrations used with permission of the Arduino Team (CC BY-NC-SA) [www.arduino.cc]. Image of sensor used courtesy of Milone Technology [www.milonetech.com].

is halved by a voltage divider network. The halving is compensated for in the Arduino sketch [www.milonetech.com].

```
//**********************************************************************
//File: LCD_Milone
//Test sketch for MKR1000 to LCD HD 44780 with
//Milone eTape sensorr connected to A2.  The sensor is
//equipped with the 0 to 5 VDC Resistance to Voltage
//Module.  The sensor's output is halved by a voltage
//divider network to keep it below 3.3 VDC.  The halving is
//compensated in software by a factor of two.
//LCD provided with MKR IoT Bundle
//**********************************************************************

#include <LiquidCrystal.h>      //LCD support library

//associate MKR1000 pins with LCD pins
const int rs = 6, en = 7, d4 = 8, d5 = 9, d6 = 10, d7 = 11;
float H20_level_float;

//initialize MKR1000 to LCD interface
LiquidCrystal lcd(rs, en, d4, d5, d6, d7);

void setup()
  {
  lcd.begin(16, 2);             //LCD's number of columns and rows
  lcd.setCursor(0, 0);          //set cursor to column 0, line 0
  lcd.print("H20 level:");      //Print message to LCD
  }

void loop()
  {
  lcd.setCursor(11, 0);         //set cursor to column 8, line 0
  H20_level_float = analogRead(A2);  //read half of sensor value
                                //convert to analog voltage and
                                //compensate for halving
  H20_level_float = H20_level_float/1023.0 * 3.3 * 2;
                                //adjust for 0 to 5 VDC equates
                                //to 0" to 12"
  H20_level_float = H20_level_float/5.0 * 12.0;
  lcd.print(H20_level_float);
```

```
  delay(1000);
  }


//***********************************************************************
```

Misting System and LED.   The greenhouse misting system is shown in Figure 4.17. A Hylaea misting system consisting of a 12 VDC fluid pump and misting delivery hardware is provided in the kit [www.mistcoolingkit.com].

An interface circuit is required between the MKR1000 and the 12 VDC fluid pump. A TIP120 Darlington transistor is used to amplify the low-level 3.3 VDC output from the MKR1000 to a 12 VDC, 1.8 A signal required to activate the pump, as shown in Figure 4.17b. A 1N4001 diode serves as protection for the inductive load. The TIP120 should be equipped with a heatsink (Jameco #326596). The 12 VDC source is supplied by the DF Robot DFR0580 Solar Power Management Module OUT 2 rated at 12 V, 8 A. An LED indicator circuit is also provided to indicate when the pump is running.

```
//***********************************************************************
//GIS_pump:  test features of the Hylaea misting system.
//    - The 12 VDC, 1.8A pump connected to MKR1000 pin D0 via
//      TIP120 Darlington transistor interface circuit.
//    - A LED is connected to pin D1 via PN2222 transistor interface
//      circuit.
//    - A potentiometer is connected to A3 to simulate the humidity
//      sensor.
//    - The input voltage to A3 is displayed on the LCD.
//***********************************************************************

#include <LiquidCrystal.h>         //LCD support library

//associate MKR1000 pins with LCD pins
const int rs = 6, en = 7, d4 = 8, d5 = 9, d6 = 10, d7 = 11;
const int pump_led = 1, mist_pump = 0;

//initialize MKR1000 to LCD interface
LiquidCrystal lcd(rs, en, d4, d5, d6, d7);

void setup()
  {
```

```
    lcd.begin(16, 2);              //LCD's number of columns and rows
    lcd.setCursor(0, 0);           //set cursor to column 0, line 0
    lcd.print("Pot voltage:");     //Print message to LCD

    pinMode(pump_led, OUTPUT);     //initialize LED pin as output
    pinMode(mist_pump,OUTPUT);     //initialize mist_pump as output
    }

void loop()
    {
    lcd.setCursor(0, 1);                //set cursor to column 0, line 1
    int pot_voltage = map(analogRead(A3),0, 1023, 0, 330);
    float pot_voltage_float = (float)(pot_voltage/100.0);
    lcd.print(pot_voltage_float);

    if (pot_voltage_float >= 2.0)   //threshold setting
       {
       digitalWrite(pump_led, HIGH);
       digitalWrite(mist_pump,HIGH);
       }
    else
       {
       digitalWrite(pump_led, LOW);
       digitalWrite(mist_pump,LOW);
       }
    delay(100);
    }

//********************************************************************
```

Vent Fan and LED.   The greenhouse vent fan system is shown in Figure 4.18. A Baosity recreational vehicle 12 VDC vent fan is mounted in the east eave of the greenhouse roof. It provides for safe venting of the greenhouse interior when it becomes too hot.

An interface circuit is required between the MKR1000 and the 12 VDC vent fan. A TIP120 Darlington transistor is used to amplify the low-level 3.3 VDC output from the MKR1000 to a 12 VDC, 1.2 A signal required to activate the fan as shown in Figure 4.18c. A 1N4001 diode serves as protection for the inductive load. The TIP120 should be equipped with a heatsink (Jameco #326596). The 12 VDC source is supplied by the DF Robot DFR0580

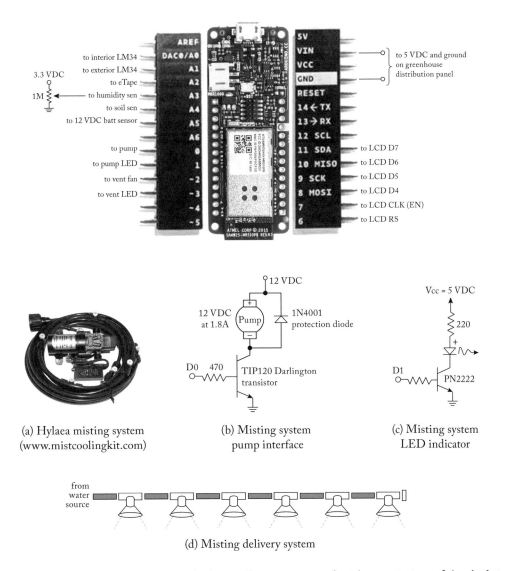

(a) Hylaea misting system
(www.mistcoolingkit.com)

(b) Misting system
pump interface

(c) Misting system
LED indicator

(d) Misting delivery system

Figure 4.17: Hylaea misting system. Arduino illustrations used with permission of the Arduino Team (CC BY-NC-SA) [www.arduino.cc].

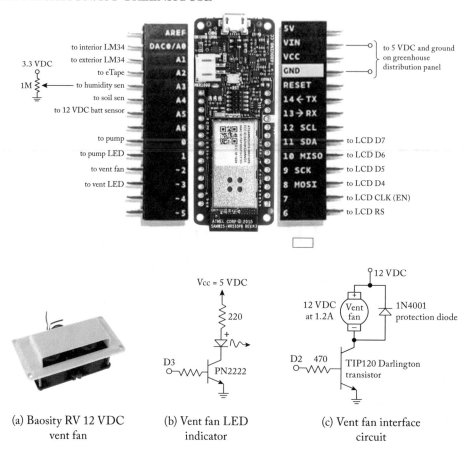

Figure 4.18: Greenhouse vent system. Arduino illustrations used with permission of the Arduino Team (CC BY-NC-SA) [www.arduino.cc].

Solar Power Management Module OUT 2 rated at 12 V, 8 A. An LED indicator circuit is also provided to indicate when the vent is running.

```
//*********************************************************************
//GIS_vent_fan:  test features of the Baosity RV 12 VDC vent fan
//   - The 12 VDC, 1.8A fan connected to MKR1000 pin D2 via
//     TIP120 Darlington transistor interface circuit.
//   - A LED is connected to pin D3 via PN2222 transistor interface
//     circuit.
//   - A potentiometer is connected to A3 to simulate the LM34 temp
```

```
//      sensor.
//   - The input voltage to A3 is displayed on the LCD.
//*************************************************************************

#include <LiquidCrystal.h>          //LCD support library

//associate MKR1000 pins with LCD pins
const int rs = 6, en = 7, d4 = 8, d5 = 9, d6 = 10, d7 = 11;
const int vent_led = 3, vent_fan = 2;

//initialize MKR1000 to LCD interface
LiquidCrystal lcd(rs, en, d4, d5, d6, d7);

void setup()
  {
  lcd.begin(16, 2);                 //LCD's number of columns and rows
  lcd.setCursor(0, 0);              //set cursor to column 0, line 0
  lcd.print("Pot voltage:");        //Print message to LCD

  pinMode(vent_led, OUTPUT);        //initialize LED pin as output
  pinMode(vent_fan,OUTPUT);         //initialize mist_pump as output
  }

void loop()
  {
  lcd.setCursor(0, 1);              //set cursor to column 0, line 1
  int pot_voltage = map(analogRead(A3),0, 1023, 0, 330);
  float pot_voltage_float = (float)(pot_voltage/100.0);
  lcd.print(pot_voltage_float);

  if (pot_voltage_float >= 2.0)   //threshold setting
    {
    digitalWrite(vent_led, HIGH);
    digitalWrite(vent_fan,HIGH);
    }
  else
    {
    digitalWrite(vent_led, LOW);
```

```
    digitalWrite(vent_fan,LOW);
    }
  delay(100);
  }


//********************************************************************
```

12 VDC Lead-Acid Battery Sensor.   To periodically check the voltage level of the 12 VDC lead-acid battery, a voltage divider network of a 10 Mohm and 1 Mohn is used. Compensation is provided in the Arduino sketch for the voltage scaling circuit.

```
//********************************************************************
//GIS_batt_check:  test features of the 12 VDC battery sensing circuit
//   - The 12 VDC lead-acid battery is provided to A5 via a
//     voltage divider circuit.
//   - The voltage divider circuit consists of a 10 Mohm and a 1 Mohm
//     resistor in series.
//   - The voltage sensed at A5 is 1/11 of the 12 VDC battery voltage
//   - The sensed value is multiplied by 11 to compensate for the
//     resistor network scaling.
//   - The input voltage to A5 is displayed on the LCD.
//********************************************************************

#include <LiquidCrystal.h>          //LCD support library

//associate MKR1000 pins with LCD pins
const int rs = 6, en = 7, d4 = 8, d5 = 9, d6 = 10, d7 = 11;

//initialize MKR1000 to LCD interface
LiquidCrystal lcd(rs, en, d4, d5, d6, d7);

void setup()
  {
  lcd.begin(16, 2);               //LCD's number of columns and rows
  lcd.setCursor(0, 0);            //set cursor to column 0, line 0
  lcd.print("Batt voltage:");     //Print message to LCD
  }
```

```
void loop()
  {
  lcd.setCursor(0, 1);                 //set cursor to column 0, line 1
  int batt_voltage = map(analogRead(A5),0, 1023, 0, 330);
  float batt_voltage_float = (float)((batt_voltage/100.0) * 11.0);
  lcd.print(batt_voltage_float);
  delay(100);
  }


//*************************************************************************
```

MKR1000 Real Time Clock.   The MKR1000 is equipped with an RTC. The RTC provides for keeping track of seconds, minutes, hours, etc. The RTC clock is configured using the Arduino RTC library. In the example below, the RTC is configured with the current time and the result is then displayed on the LCD.

```
//*************************************************************************
//MKR1000_RTC: Real Time Clock sketch for MKR1000 using onboard RTC
//This example code is in the public domain
//   http://arduino.cc/en/Tutorial/SimpleRTC
//   created by: Arturo Guadalupi <a.guadalupi@arduino.cc> 15 Jun 2015
//   modified by Andrea Richetta <a.richetta@arduino.cc>   24 Aug 2016
//   modifed by S. Barrett Oct 20, 2020 - RTC to LCD
//      - test sketch for MKR1000 to LCD HD 44780
//*************************************************************************

#include <RTCZero.h>                //RTC support library
#include <LiquidCrystal.h>          //LCD support library


RTCZero rtc;                            //Create an rtc object
                                        //current time and date
const byte seconds = 0;  const byte minutes = 30; const byte hours = 4;
const byte day = 20;     const byte month = 10;   const byte year = 20;

//associate MKR1000 pins with LCD pins
const int rs = 6, en = 7, d4 = 8, d5 = 9, d6 = 10, d7 = 11;
```

```
//initialize MKR1000 to LCD interface
LiquidCrystal lcd(rs, en, d4, d5, d6, d7);

void setup()
{
Serial.begin(9600);                //serial monitor setup
rtc.begin();                       //initialize Real Time Clock
rtc.setHours(hours);               //set current time
rtc.setMinutes(minutes);
rtc.setSeconds(seconds);
rtc.setDay(day);                   //set current date
rtc.setMonth(month);
rtc.setYear(year);

lcd.begin(16, 2);                  //LCD's number of columns and rows
lcd.setCursor(0, 0);               //set cursor to column 0, line 0
lcd.print("Time:");                //Print message to LCD
}

void loop()
{
print2digits(rtc.getDay());
Serial.print("/");
print2digits(rtc.getMonth());
Serial.print("/");
print2digits(rtc.getYear());
Serial.print(" ");
print2digits(rtc.getHours());
Serial.print(":");
print2digits(rtc.getMinutes());
Serial.print(":");
print2digits(rtc.getSeconds());
Serial.println();

lcd.setCursor(0, 1);               //set cursor to column 0, line 1
lcd.print(rtc.getDay());
lcd.print("/");
lcd.print(rtc.getMonth());
```

```
lcd.print("/");
lcd.print(rtc.getYear());
lcd.print(" ");
lcd.print(rtc.getHours());
lcd.print(":");
lcd.print(rtc.getMinutes());
lcd.print(":");
lcd.print(rtc.getSeconds());

delay(1000);
}


//***********************************************************************

void print2digits(int number)
{
if(number < 10)
  {
  Serial.print("0"); // print a 0 before if the number is < than 10
  }

Serial.print(number);
}


//***********************************************************************
```

GIS System Code. The UML activity diagram for the GIS system code is provided in Figure 4.19. The corresponding code is not provided. The pieces to assemble the code have been provided throughout this chapter. Selected pieces may be chosen for your specific application.

GIS Printed Circuit Board. The components for the GIS System are mounted to a printed circuit board (PCB), as shown in Figure 4.20.

Enclosure. The completed GIS System is mounted within a QILIPSU hinged cover, stainless-steel latch, junction box with mounting plate. The layout of the junction box is provided in Figure 4.21 and the final result is provided in Figure 4.22.

Testing. The final project step is to thoroughly test all system features. A test plan is developed to test and document the proper operation of each system feature and the overall system. The

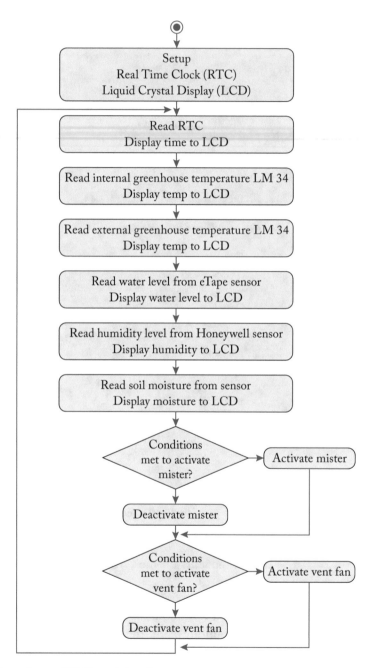

Figure 4.19: Greenhouse UML activity diagram.

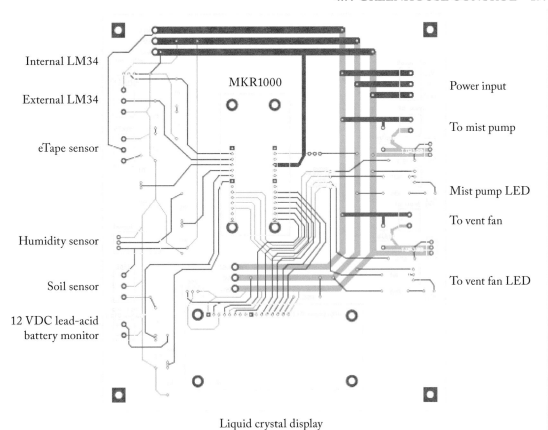

Figure 4.20: Greenhouse printed circuit board.

complete greenhouse is shown in Figure 4.23. The GIS system with weather station is shown in Figure 4.24.

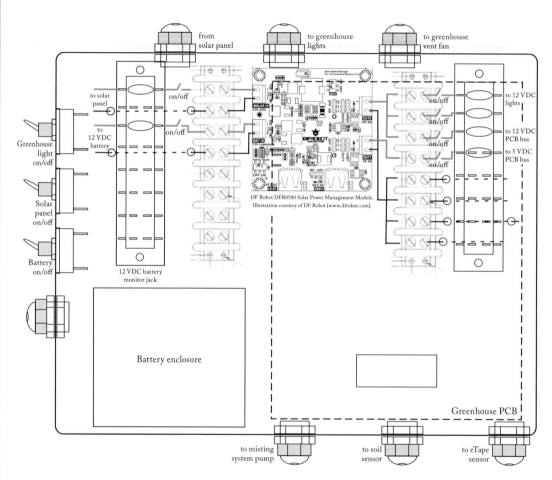

Figure 4.21: **GIS panel layout.**

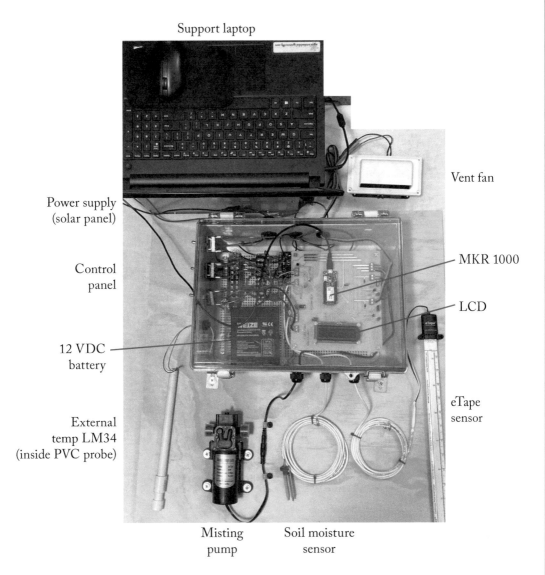

Figure 4.22: **GIS** system.

Figure 4.23: Greenhouse.

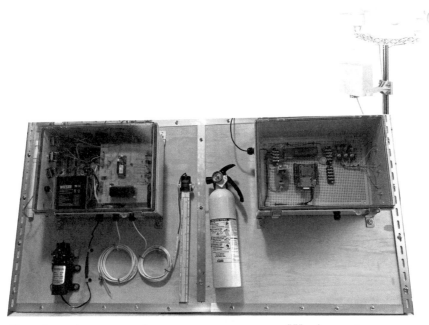

Greenhouse instrumentation system   Weather station

Figure 4.24: Greenhouse instrumentation systems.

4.8   APPLICATION: EXPLORATION WITH MKR IoT BUNDLE

In this exercise we use the Arduino MKR IoT Bundle to connect an Arduino MKR 1000 device to the internet. The MKR 1000 features a 32-bit ATSAMD21 processor operating at 48 MHz. Although the processing board is physically small, it is full featured with 256 KB of flash and 32 KB of RAM. The processor hosts SPI, I2C, and UART communication systems. The processor is also equipped with pulse width modulation (PWM) features and multiple channels of analog-to-digital conversion channels. The pinout for the MKR 1000 is provided in Figure 4.25. There are several methods of providing battery power to the MKR 1000 including an input connector for a 3.7 VDC rechargeable LiPo (lithium-ion polymer) battery.

The MKR 1000 is actually a system on a chip. It uses the Microchip [www.microchip] ATSAM25 system on a chip with three onboard systems:

- SAMD 21 Cortex–M0+ 32-bit low power ARM processor,

- WINC1500 low power 2.4 GHz WiFi module and antenna, and

- ECC 508 crypto authentication unit.

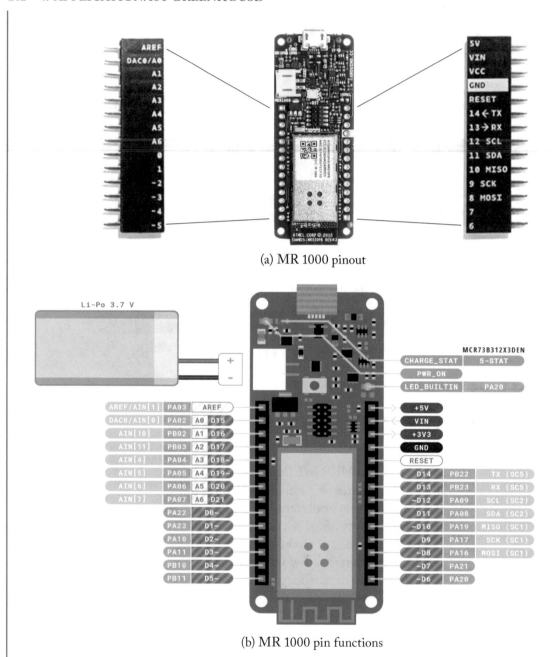

(a) MR 1000 pinout

(b) MR 1000 pin functions

Figure 4.25: **MKR 1000 pinout.** Illustrations used courtesy of Arduino CC-BY-SA [www.arduino.cc].

The Arduino MKR IoT Bundle also includes a prototyping breadboard, battery clip, and multiple external components to implement a variety of different projects that interact with the internet.

Objective.   Using the Arduino MKR 1000 WiFi board, an LED will be controlled via a remote computer. This is accomplished using the Arduino IoT cloud interface.

Arduino IoT Interface.   The Arduino IoT interface is shown in Figure 4.26.

The Arduino company has developed an extensive infrastructure to allow for the streamlined development of IoT applications. Arduino has made this freely available to IoT enthusiasts. To gain access to these IoT assets, an account may be established at: https://create.arduino.cc/iot/things.

In addition to software assets, Arduino has provided several excellent tutorials on getting started with the IoT Cloud.

- "Arduino MKR1000 Getting Started" by Dr. C. Mahmoudi available at https://create.arduino.cc/projecthub/133030/iot-cloud-getting-started-c93255. This tutorial provides a detailed step-by-step approach to attaching and controlling several "things" via the cloud including an LED, a switch, and a potentiometer. These serve as representative examples for a digital output device (LED), digital input device (switch), and an analog input device (potentiometer).

- "Getting started with the Arduino IoT Cloud" available at https://www.arduino.cc/en/Tutorial/iot-cloud-getting-started.

In this application, we summarize the steps from the first tutorial. We encourage you to follow the detailed steps provided in the tutorial.

Summary of Steps.

1. Download the latest version of Arduino IDE from https://www.arduino.cc/en/Main/Software to your laptop or PC.

2. Connect the MKR 1000 to your computer via the cable provided in the Bundle.

3. Install the drivers for the MKR 1000 from within the Arduino IDE. Go to Tools − > Board − > Board Manager − > Arduino SAMD Boards (32-bits ARM Cortex - M0+)

4. Log into the Arduino IoT Cloud website at: https://create.arduino.cc/iot/things. You will be led through a series of steps to configure an LED hardware interface and control the LED via a remote computer. Provided below is an overview of the steps followed to develop the IoT application.

   - Connect the LED interface to the MKR 1000 board.

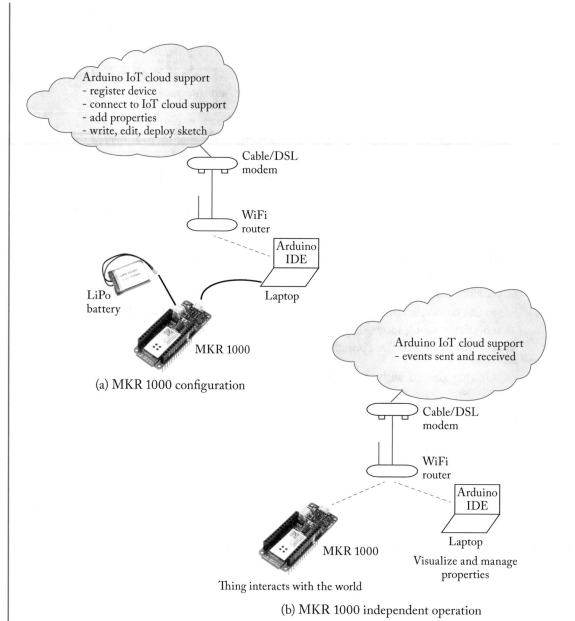

(a) MKR 1000 configuration

(b) MKR 1000 independent operation

Figure 4.26: MKR 1000 configuration. MKR 1000 illustrations used courtesy of Arduino CC-BY-SA [www.arduino.cc].

- Add the MKR 1000 to the Arduino IoT Cloud as a **Thing**.
- Provide the **Thing** a set of **Properties**. The properties represent the hardware components connected to the MKR 1000. In this example this will be the LED.
- Set the Property Permissions.
- The sketch (main.ino) to support the IoT interface is automatically generated with three supporting files: ReadMe.adoc, thingProperties.h, and Secret.
- At the Secret tab, insert the name and password for the WiFi network supporting the MKR 1000 board.
- Upload the generated sketch to the MKR 1000 board.

5. The MKR 1000 will connect to the Arduino IoT Cloud and begin communicating with it. The interaction of the MKR 1000 with the Arduino IoT Cloud may be observed on the Arduino IDE's Serial Monitor.

6. You can control the operation of the LED from your computer by selecting "GO TO IOT CLOUD." This will take you to the Thing's page on the Arduino IoT Cloud. You can control the LED via the dashboard button.

7. Complete the tutorial for the switch and the potentiometer.

Extension.   Using lessons learned from the tutorial, equip the MKR1000 with an LM34 Precision Fahrenheit Temperature Sensor. Monitor the temperature from the cloud.

4.9   SUMMARY

The goal of this chapter was to demonstrate in action the concepts discussed in this book. Simply put, our goal is to provide the theory, design, and construction of a passively heated greenhouse. We equip the greenhouse with instrumentation to monitor and control key parameters. Using IoT concepts key parameters will be monitored and controlled via a remote computer.

4.10   REFERENCES

[1] Arduino homepage. www.arduino.cc

[2] Baird, C. (2011). *The Complete Guide to Building Your Own Greenhouse*, Atlantic Publishing.

[3] Doxon, L. *How to Calculate Greenhouse Heating*. https://homeguides.sfgate.com

[4] Encinias, V. *Efficient Greenhouse Design*. https://gpnmag.com

[5] *Five Low-Tech Winter Greenhouse Heating Techniques*. https://www.rimolgreenhouses.com

[6] Hanes, D., Salgueiro, G., Grossetete, P., Barton, R., and Henry, J. (2017). *IoT Fundamentals—Networking Technologies, Protocols, and Use Cases for the Internet of Things*, Cisco Press.

[7] Honeywell, *HIH-4030/31 Series Humidity Sensors*, Honeywell Sensing and Control. www.honeywell.com/sensing

[8] Kelley, A. and Pohl, I. (1998). *A Book on C—Programming in C*, 4th ed., Addison Wesley.

[9] Lindsey, C. and Plinke, M. (2016). *The Year-Round Solar Greenhouse*, New Society Publishers.

[10] Marshall, R. (2006). *How to Build Your Own Greenhouse*, Storey Publishing.

[11] Milone Technologies, *0–5 VDC Linear Resistance to Voltage Module PN–05V00199 Rev 2*. www.milonetech.com

[12] Milone Technologies, *eTape Continuous Fluid Level Sensor PN–12110215TC–X*. www.milonetech.com

[13] National Semiconductor, *LM34 Precision Fahrenheit Temperature Sensors*, DS006685, National Semiconductor Corporation, www.national.com, 2000.

[14] Oehler, M. (2007). *The Earth-Sheltered Solar Greenhouse Book*, Mole Publishing Company.

[15] Schmidt, P. (2011). *The Complete Guide to Greenhouses and Garden Projects*, Creative Publishing International.

[16] *Passive Solar Greenhouse*, Bradford Research Center, University of Missouri, 2017. http://bradford.cafnr.org/passive-solar-greenhouse

[17] Schiller, L. *Three Methods for Heating Greenhouses for Free*. https://www.motherearthnews.com

[18] Schiller, L. *How to Design a Year-Round Solar Greenhouse*. https://www.motherearthnews.com

[19] Thoma, M. *Seven Useful Features You Need in a Passive Solar Greenhouse*. tranquilurbanhomestead.com

4.11   CHAPTER PROBLEMS

4.1. Provide a design to allow the LM34 to sense negative temperatures.

4.2. In the GIS system, voltage divider circuits were used with several of the sensors. Why? Explain in detail.

4.3. Provide a structure chart for the GIS system code.

4.4. Provide the overall GIS system code.

4.5. Develop the overall system software for the GIS System.

4.6. Develop a test plan to insure requirements have been met for the GIS system.

4.7. What are the critical variables that should be regularly monitored?

4.8. If the greenhouse is within 10 meters of your home, design a system to report critical variables to your home PC. Explain in detail your choice of technology.

4.9. If the greenhouse is within 10 meters of your home, design a system to report critical variables to your cell phone. Explain in detail your choice of technology.

4.10. If the greenhouse is within 50 meters of your home, design a system to report critical variables to your home PC. Explain in detail your choice of technology.

4.11. Design a system to log and plot the internal and external greenhouse temperature every hour.

A P P E N D I X   A

Programming the ATmega328

There are several different methods of programming the ATmega328.

1. **Onboard Arduino UNO R3 Programming:** In Chapter 1, we described how to program the ATmega328 while onboard the Arduino UNO R3 using the Arduino IDE. This technique allows access to the user-friendly features of the Arduino IDE and also retains the Arduino operating environment.

2. **Onboard Arduino UNO R3 using In System Programming (ISP):** This technique employs an ISP programmer (e.g., AVR Dragon) to program the ATmega328 while onboard the Arduino UNO R3. This technique allows access to the user-friendly hardware interface features of the Arduino UNO R3. However, when a program is loaded to the ATmega328 via the ISP programmer, the Arduino operating environment is overwritten. It is important to note that the Arduino operating environment can be rewritten to the ATmega328 if need be.

3. **ATmega328 using ISP:** This technique employs an ISP programmer (e.g., AVR Dragon) to program the ATmega328 via the Serial Peripheral Interface (SPI) system.

    In the next section we describe how to program the ATmega328 using ISP techniques.

A.1   ISP HARDWARE AND SOFTWARE TOOLS

Programming the ATmega328 requires several hardware and software tools. For software tools, a compiler and device programming support is required. Throughout the book we provide examples using both the ImageCraft JumpStart C for AVR compiler www.imagecraft.com and also the Atmel Studio gcc compiler www.atmel.com. We use the Atmel Studio software suite with the AVR Dragon programmer to download and program the microcontroller. The connection between the host computer and the AVR Dragon is shown in Figure A.1 with detailed instructions provided below.

A.2   ImageCraft JumpStart C FOR AVR COMPILER DOWNLOAD, INSTALLATION, AND ATmega328 PROGRAMMING

Throughout the text, we provide examples using the ImageCraft JumpStart C for AVR compiler. This is an excellent, well-supported, user-friendly compiler. The compiler is available for

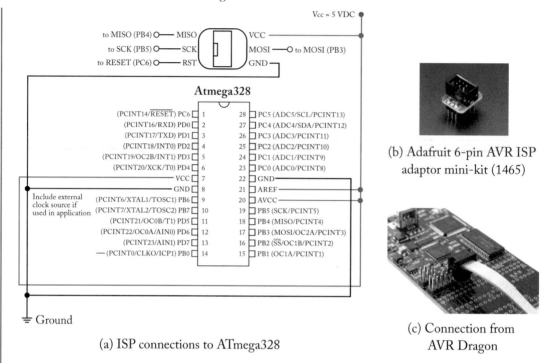

(b) Adafruit 6-pin AVR ISP adaptor mini-kit (1465)

(c) Connection from AVR Dragon

(a) ISP connections to ATmega328

Figure A.1: Programming the ATmega328 with the AVR Dragon. (Image of Adafruit 1465 used with permission of Adafruit [www.adafruit.com]. Microchip AVR Dragon illustration used with permission of Microchip, Incorporated [www.microchip.com].)

purchase and download at www.imagecraft.com. Details on compiler download and configuration are provided there. ImageCraft allows a 45-day compiler "test drive" before securing a software license. One of the authors (sfb) has used variants of this compiler for over a decade on multiple Microchip AVR products. You can expect prompt, courteous service from the company. There are other excellent compilers available. The compiler is used to translate the source file(s) (testbench.c and testbench.h) into machine language (testbench.hex) for loading into the ATmega328. We use Microchip's Atmel Studio to load the machine code into the ATmega328.

1. Download and install ImageCraft JumpStart C for AVR compiler.

2. Create a new project (File − > New − > Project) and select "ImageCraft AVR Project." Press Next.

3. Provide a descriptive project title and browse to the desired folder location.

4. Select target as ATmega328. Details of the microcontroller and connected programming pod (e.g., AVR Dragon) will populate drop down windows. Press OK. A project with a "main.c" template will be created.

5. Open "main.c" and write your program.

6. With program writing complete, build project (Build − > Build). Correct any syntax errors and rebuild project. Repeat this process until no syntax errors remain. Upon a successful program build, a filename.hex and filename.elf are created.

7. Reference the next section to program the ATmega328 using ISP techniques.

A.3   ATMEL STUDIO DOWNLOAD, INSTALLATION, AND ATmega328 PROGRAMMING

1. Download and install the latest version of Atmel Studio.

2. Connect the AVR Dragon to the host PC via a USB cable. The AVR Dragon drivers should automatically install.

3. Configure hardware:

   • Configure the Adafruit 6-pin AVR ISP adaptor mini-kit (1465) as shown in Figure A.1.

   • Connect the Adafruit 6-pin AVR ISP adaptor to the AVR Dragon via 6-pin socket IDC cable.

   • Connect the Adafruit 6-pin AVR ISP adaptor to ATmega328 target as shown.

4. Program Compiling and Programming

   • Start Atmel Studio.

   • If using the gcc compiler: File − > New − > Project − > GCC Executable Project − > <filename>

   • If using the gcc compiler: Write program.

   • If using the gcc compiler: Build program.

   • Tools Device Programming

     − Dragon
     − ATmega328
     − Interface: ISP Apply
     − Insure target chip has 5 VDC applied

   Apply

- Read signature, read target voltage

- Memories: Flash: Program–Browse for desired "filename.hex" and press "Program."

- Fuses: Ext. Crystal Osc. 8.0- MHz: Program

Author's Biography

STEVEN F. BARRETT

Steven F. Barrett, Ph.D., P.E., received a B.S. in Electronic Engineering Technology from the University of Nebraska at Omaha in 1979, an M.E.E.E. from the University of Idaho at Moscow in 1986, and a Ph.D. from The University of Texasat Austin in 1993. He was formally an active duty faculty member at the United States Air Force Academy, Colorado and is now the Associate Vice Provost of Undergraduate Education at the University of Wyoming and Professor of Electrical and Computer Engineering. He is a member of IEEE (senior) and Tau Beta Pi (chief faculty advisor). His research interests include digital and analog image processing, computer-assisted laser surgery, and embedded controller systems. He is a registered Professional Engineer in Wyoming and Colorado. He co-wrote with Dr. Daniel Pack several textbooks on microcontrollers and embedded systems. In 2004, Barrett was named "Wyoming Professor of the Year" by the Carnegie Foundation for the Advancement of Teaching and in 2008 was the recipient of the National Society of Professional Engineers (NSPE) Professional Engineers in Higher Education, Engineering Education Excellence Award.

Index

Printed in the United States
by Baker & Taylor Publisher Services